Permeable Reactive Barriers

Permeable Reactive Barriers

Edited by

GENEVIEVE A. BOSHOFF
Mouchel Parkman, Ellesmere Port, Cheshire, UK

BRIAN D. BONE
Environment Agency Science Group, Solihull, West Midlands, UK

IAHS Publication 298
in the IAHS Series of Proceedings and Reports

Published by the International Association of Hydrological Sciences 2005

IAHS Publication 298
ISBN 1-901502-23-6

British Library Cataloguing-in-Publication Data.
A catalogue record for this book is available from the British Library.

The papers included in this volume were peer reviewed and revised, prior to publication.

IAHS is indebted to the employers of the Editors for the invaluable support and services provided that enabled them to carry out their task effectively and efficiently.

Publications in the series of Proceedings and Reports are available from:
IAHS Press, Centre for Ecology and Hydrology, Wallingford, Oxfordshire OX10 8BB, UK
tel.: +44 1491 692442; fax: +44 1491 692448; e-mail: jilly@iahs.demon.co.uk

Printed in The Netherlands by Krips BV, Meppel.

Foreword

Prior to the early 1980s, there was a common view that once contaminated, an aquifer was lost as a water resource for the foreseeable future. In response to highly publicized instances of groundwater contamination, such as the Love Canal, and a growing awareness of the high incidence of chlorinated organic chemicals in groundwater supplies, in 1980 the US Congress passed the Comprehensive Environmental Response, Compensation and Liability Act (CERCLA, or better known as Superfund), and in 1984, amended the Resource Conservation and Recovery Act (RCRA), in both cases, legislating the cleanup of contaminated groundwater. A new industry was spawned: groundwater remediation.

With little experience and technical background, it is not surprising that hydrologists and engineers brought their traditional knowledge to bear on this new challenge. In particular, the primary method of groundwater remediation was to pump the contaminated water to the ground surface, remove the contaminants by some conventional treatment process and discharge the water (pump-and-treat). By the late 1980s, with few sites cleaned up, much longer times required for cleanup than predicted and escalating costs, one might have expected a radically different approach to remediation to be welcomed with open arms.

The use of granular iron, in a permeable reactive barrier (PRB) configuration, for removal of chlorinated organic contaminants, was first proposed in the early 1990s. The initial submission to a refereed journal was rejected and the proposed approach was generally greeted with a high degree of suspicion and scepticism. Nevertheless, through the interest of the scientific community and the results of early field demonstrations, by the mid 1990s, the technology had gained considerable credibility. Currently more than 100 PRBs containing granular iron have been installed for removal of chlorinated organic contaminants from groundwater, and perhaps an equal number containing other media have been installed for removing other contaminants. These numbers are commonly cited as evidence of the great success and acceptance of the PRB technology. On the other hand, relative to the magnitude of the problem, with contaminated sites numbering many hundreds of thousands, the influence of the technology on global environmental quality is, as yet, little more than miniscule.

The contributions in this volume, and the conference from which they were derived, provide considerable insight concerning the evolution of the PRB technology. First, it is clear that the technology has gained global interest and acceptance. Perhaps of greater importance, the nature of the questions being addressed has evolved from "if" to "how". Indeed, the reactions of iron with chlorinated solvents are generally accepted and attention has now turned to more effective means of applying the technology. Active areas of research reflected in this volume include alternative reactants for other types of contaminants, configurations and reactants suitable for mixed contaminant plumes, emplacement alternatives and long-term performances. With the maturation of the technology, as reflected in these topics, it appears certain that PRBs will continue as a significant remediation technology and that their stature and acceptance is likely to increase in the future.

R. W. (Bob) Gillham
University of Waterloo
Ontario, Canada

Preface

Groundwater pollution is a global problem, impacting on the quality of life of people in both developed and developing countries. If we are to meet the ever-growing challenge to remediate these natural resources, then we need to look beyond the traditional methods of cleanup and embrace "alternative" technologies. One such technology is permeable reactive barriers (PRBs), which have had much success in the remediation of chlorinated solvent contaminated groundwaters and are now gaining recognition for the treatment of a range of other contaminants, both inorganic and organic.

This publication comprises a selection of peer-reviewed work presented at the first International Symposium on Permeable Reactive Barriers, held in Belfast, Northern Ireland in March 2004. This was the first international symposium of its kind focusing solely on this technology and providing researchers, industry and regulators with a platform from which to develop ideas and foster collaborations. Over 100 delegates attended the conference, representing both developed and developing countries, and attesting to the growing interest in this technology.

The papers focus on innovations in the field and the long-term performance of PRBs, and include the presentation of performance data from the first commercial granular iron barrier installed. I hope that after reading these contributions you will continue to support this technology and its many success stories.

Genevieve A. Boshoff

Mouchel Parkman, Ellesmere Port,
Cheshire, UK

Contents

1 Long-term Performance of PRBs

Long-term performance of Fe^0-PRBs to treat arsenic and 1,2-DCE(*cis*) under different geochemical conditions

RALF KÖBER[1], FRANCESCA GIAROLLI[2], MARKUS EBERT[1] & ANDREAS DAHMKE[1]

1 *Institute of Geosciences, University of Kiel, Ohlshausenstrasse 40,D-24118 Kiel, Germany*
rkoeber@gpi.uni-kiel.de

2 *Department of Chemistry, University of Rome "La Sapienza", P.le Aldo Moro 5, I-00185 Rome, Italy*

Abstract Fe^0 has been previously shown to have potential for the removal of arsenic (As) from groundwater. Several sets of coupled experimental columns were run to investigate if As can be removed together with 1,2-DCE(cis) by Fe^0-PRBs under different geochemical situations. As was removed with half-life times of approximately 2 h to concentrations of mostly below 10 μg l^{-1} throughout the whole experiment under oxidic and reduced conditions. In contrast, 1,2-DCE(*cis*) dechlorination decreased with time, independently of the input redox conditions. The reason for the inhibited dechlorination is assumed to be passivation by Si adsorption following the onset of microbial sulphide reduction (MSR). Precipitation of arsenic sulphides and sorption of As to iron sulphides seems to stabilize As removal.

Key words 1,2-DCE(cis); arsenic; inhibition; long-term performance; microbial sulphide reduction; permeable reactive barrier; XAS; VOC

INTRODUCTION

Groundwater contamination by arsenic is considered to be one of the most important environmental reasons for worldwide cancer mortality. An increasing number of publications showed that permeable reactive barriers composed of zero-valent iron (Fe^0-PRBs) could be a promising remediation approach since corrosion products of Fe^0 adsorb arsenic and As can co-precipitate with the forming iron (hydr-)oxides (Lackovic *et al.*, 2000; Manning *et al.*, 2002; Melitas *et al.*, 2002a,b; Su & Puls, 2001a,b; Vlassopoulos *et al.*, 2002).

Although there has been a series of investigations concerning the association of As with corrosion products, redox behaviour and the influence of common anions in groundwater, there are still some questions regarding the suitability of Fe^0 for *in situ* application for periods of years or decades. Evaluation of the competitive effects of inorganic anions on As removal by Fe^0 showed the strongest inhibition for silicate and phosphate (Su & Puls, 2001a, 2003). Sulphide can reduce the formation of new iron (hydr-)oxide adsorption sites by decreasing available Fe^{2+} and Fe^{3+} due to iron sulphide precipitation, but otherwise iron sulphides can also remove As by incorporation or adsorption. Arsenic can further precipitate as arsenic sulphides like orpiment (As_2S_3) or realgar (AsS). Despite this diverse possibility for sulphide of affecting As removal, thc influence of sulphide has not yet been studied in further detail.

Most of the solid phase analysis, to investigate the bonding of As to the corrosion products, was performed with materials from laboratory experiments without indications of microbial activity (Farrell *et al.*, 2001; Manning *et al.*, 2002; Melitas *et al.*, 2002b; Su & Puls, 2001b). Those studies showed that adsorption of As to iron (hydr-)oxides is the most important process for the As removal by Fe^0. Otherwise on-site column tests under more realistic conditions showed association of As with sulphur (Nikolaidis *et al.*, 2003). This is another indication that sulphate or sulphide could play an important role in As removal by Fe^0.

So far no long-term investigations into the As removal by Fe^0-PRBs has been performed. Column studies and on-site experiments with considerably lower retention times than typical for most PRBs all showed As breakthrough after relatively short times of weeks or a few months (Nikolaidis *et al.*, 2003; Su & Puls, 2003). Nevertheless, As breakthrough was mostly incomplete and the outflow stabilized on a fraction of the input solution As concentration, indicating continuous formation of adsorption sites (Su & Puls, 2003). Although these findings are encouraging, it is unknown whether Fe^0 can be used *in situ* to remove As to levels less than 50 or 10 $\mu g\ l^{-1}$ for periods of years or decades.

In contrast to the removal of As by Fe^0, several studies have investigated the longevity of Fe^0 for the degradation of volatile organic compounds (VOC) and showed that silica adsorption and carbonate precipitation can be two important processes with the potential to decrease VOC degradation over time (e.g. Klausen *et al.*, 2003). Microbial sulphide reduction (MSR) can affect VOC degradation by Fe^0 in multiple ways, but only a few studies focused on this topic (e.g. Köber *et al.*, 2002).

Against this background, the main focus of this investigation is to test whether low As concentrations can be maintained under different geochemical conditions in the long-term, and if sulphide affects the longevity of Fe^0 to remove As or VOC. The experimental setup, described below, was additionally used to investigate if the mobility of As can be reduced or increased within contaminated sediment using upstream PRBs. The preliminary results from the upstream study are reported elsewhere (Köber *et al.*, 2003, 2004).

METHODS

Four sets of column experiments (Ø: 10 cm, L: 100 cm), respectively, composed of two or three sequential columns connected by viton tubing were performed using artificial groundwater (Fig. 1). Artificial groundwater was made by adding salts and acid to tap water to achieve approximately a pH value of 7.5, concentrations (in mg l^{-1}) of 40 Na, 4 K, 150 Ca, 20 Mg, 13 Si, 70 Cl, 23 NO_3^-, 140 SO_4^{2-} and an alkalinity of 5 mmol l^{-1}. Before this solution was fed to the columns, a 1,2-DCE(*cis*) solution was added in a mixing chamber. The sequence in column sets 2–4 starts with a column containing the upstream PRB materials (organic matter (compost, chaff and wood chips), gypsum, a mixture of both) followed by a column filled with the As contaminated sediment from a contaminated site in Wiesbaden (Germany) and ends with a Fe^0 column (sponge iron, ISPAT, Hamburg, Germany).

As, Fe, Mn, Ca, Mg, K, Na and Si were analysed by ICP-OES, SO_4^{2-}, NO_3 and Cl^- by ion chromatography with conductivity detection, and chlorinated ethenes by GC–ECD

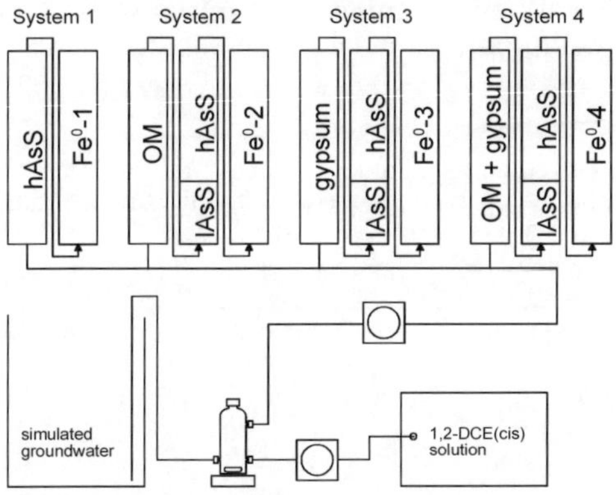

Fig. 1 Setup of sequential column experiments (hAsS, sediment with high As content; lAsS, sediment with low As content; OM, organic matter).

and GC–FID. Concentrations of As lower than 10 μg l^{-1} in samples from Fe^0 columns were measured by ICP–MS. Sulphide was analysed by photometry and alkalinity was measured by HCl titration. For the measurement of pH and E_H electrodes were used. Additionally to the standard analyses, some samples were analysed for the As species by IC–ICP–MS at the UFZ–Umweltforschungszentrum Halle-Leipzig GmbH. Solid phase investigations are still under way, but X-ray absorption spectroscopy (XAS) was used to investigate the bonding between As and its neighbour atoms.

RESULTS AND DISCUSSION

Outflow solutions from the sediment columns which were fed to the Fe^0 columns exhibited diverse chemical compositions with differing redox conditions, differing As species and concentrations, differing salt content and changing 1,2-DCE(*cis*) concentrations (Table 1). Independently of the geochemical conditions produced by the different upstream materials and sediments, concentrations of As were decreased by Fe^0 and its corrosion products to below 1–10 μg l^{-1}. Initially pH increased fast to 11–12 within the first few centimetres and remained constant to the outflow. With elapsing time pH increase slowed down and maximum pH reduced to 9–10. These pH changes were generally observed in all Fe^0-filled columns and are shown for Fe^0-1 in Fig. 2. This trend of slower and lower hydroxide release could be a result of decreased corrosion rates caused by ongoing precipitate formation on Fe^0 surfaces. In the last third of the column lengths As concentrations can be considered as constant and in equilibrium with the solid phase. It is assumed that iron (hydr-)oxides are the As concentration limiting phase in this section of the column since changes in alkalinity and sulphide concentrations are negligible in this area. Si which also adsorbs to iron (hydr-)oxides is removed too, but to a lower extent than As (Fig. 3). As, and particularly Si concentrations, showed a significant correlation with pH at a flow path of 77 cm (Fig. 4). As(V) and As(III) adsorption by several iron (hydr-)oxides had been shown to decrease with increasing pH above pH 9 (Raven *et al.*, 1998; Dixit & Hering, 2003). Decreasing As and Si outflow concentrations are therefore interpreted as a result of the decreasing outflow pH in all Fe^0-filled columns from initially 11–12 to finally 9–10 within the one-year period of examination.

Table 1 Average input solutions of Fe^0 columns. (Values separated by slash denote [before upstream material setup (reference conditions)/after upstream material set up]. Arrows indicate a continuous increase or decrease. Numbers in brackets are most representative values.

	Fe^0–1	Fe^0–2	Fe^0–3	Fe^0–4
Upstream material:	Without (reference)	Gypsum	Organic matter	Organic matter + gypsum
pH	7.3–7.8	7.2–7.5	6.5–7.9	7.0–7.6
E_H (mV)	110–180	50–200	130 → –80 (50)	200 → –140 (–110)
As_{tot} (mg l^{-1})	3–6.5	3–5.5	6 / 8–200	7 / 14–200
As(III) (%)	<3	<3	<3 / >95	<3 / >90
As(V) (%)	>97	>97	>97 / <5	>97 / <10
Fe (mg l^{-1})	<0.1	<0.1	<0.1 / 1–80	<0.1 / <0.1–23
Mn (mg l^{-1})	<0.2	<0.4	<0.2 / 0.8–9 (1)	<0.1 / 9 → 0.5
Si (mg l^{-1})	10–14	11–16	11 / 17	11 / 15
Ca (mg l^{-1})	150	150 / 750–1050	150	150 / 700–1050
Mg (mg l^{-1})	16–22	18–23	12–22	18–24
K (mg l^{-1})	4	4	4	4
Na (mg l^{-1})	40	40	40	40
S (–II) (mg l^{-1})	<0.1	<0.1	<0.1 → 2	<0.1 / 0.2 → 130 (50)
SO_4^{2-} (mg l^{-1})	140	140 / 1500	140 / <5	140 / 1250
NO_3^- (mg l^{-1})	21	21	21 / 0	21 / 0
PO_4^{3-} (mg l^{-1})	<1	<1	<1	<1
Cl^- (mg l^{-1})	70	70	70	70
Alkalinity (mmol l^{-1})	5	5	5 / 10	5 / 9
1,2–DCE(cis) (mg l^{-1})	0.004–1.5	0.005–900	0.018–8	0.015–7
NH_4^+ (mg l^{-1})	<1	<1	<1	<1
CH_4 (mg l^{-1})	<0.01	<0.01	<0.01 / 20	<0.01 / 4 → 0.2
Characteristics	oxidic	oxidic, high Ca + SO_4^{2-}	reduced, temporary huge As	reduced, high Ca + SO_4^{2-} temporary huge As

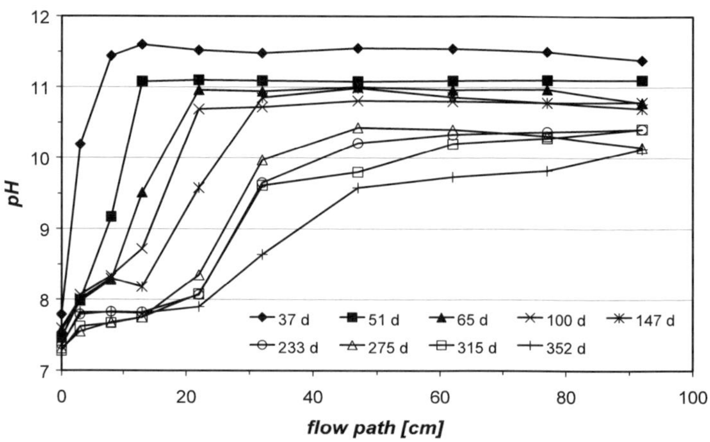

Fig. 2 pH in Fe^0-1.

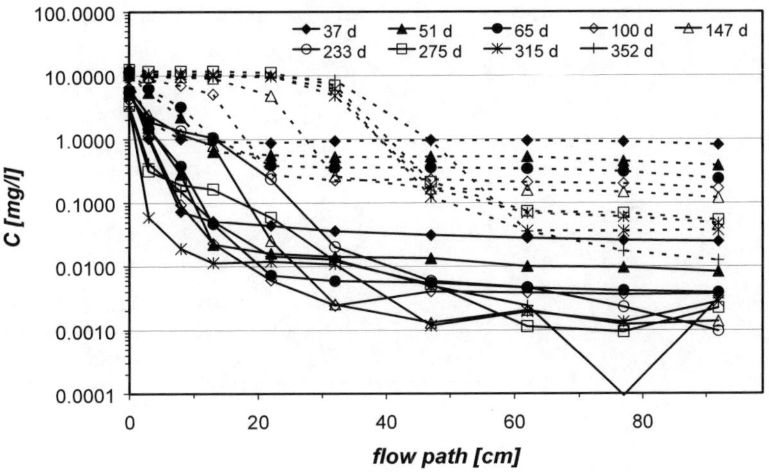

Fig. 3 Arsenic (solid lines) and silica (dashed lines) concentrations in Fe^0-1.

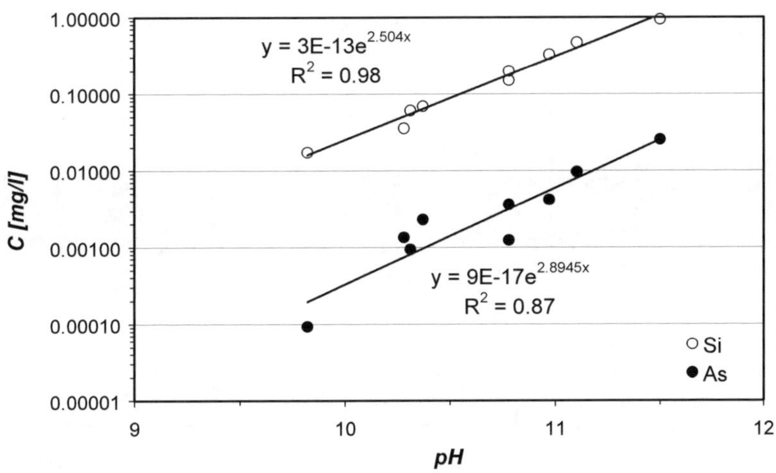

Fig. 4 pH dependency of arsenic and silica concentrations in Fe^0-1 near the outflow (77 cm).

As(III) and As(V) had to be simultaneously adsorbed by Fe^0-2 and Fe^0-4, since more than 90% of the As which dissolved from the preset sediments was As(III) and As(V) was the bulk of the remainder (Table 1). As(III) concentrations decreased faster than As(V) concentrations and As(III) in general decreased to lower concentrations than As(V) at the outflow despite higher As(III) input concentrations (Fig. 5). Thus As(V) seems to be the limiting species when mixed As(III)/As(V) contaminations are treated with Fe^0.

Removal kinetics were calculated to evaluate the longevity of Fe^0 for As treatment. It was previously shown that As removal by Fe^0 is best described by a mixed kinetic model for relatively simple solution compositions and for relatively short periods in the range of hours to days (Farrell *et al.*, 2001; Melitas *et al.*, 2002b). Possibly due

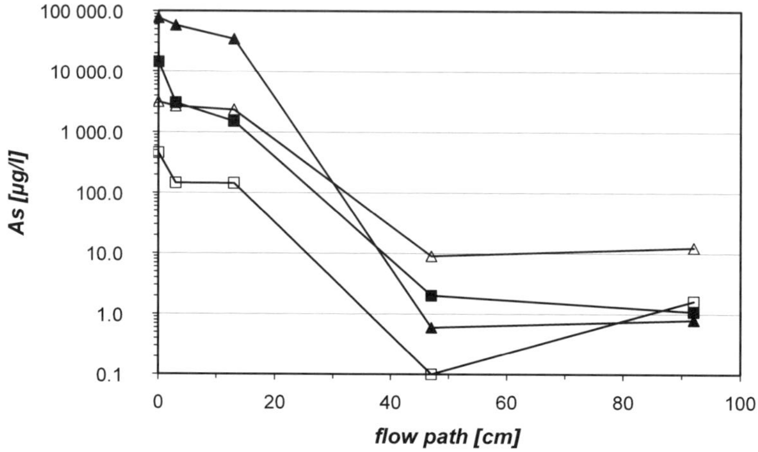

Fig. 5 As(V) (open) and As(III) (bold) concentrations after 142 (triangles) and 197 (squares) days in Fe^0-2.

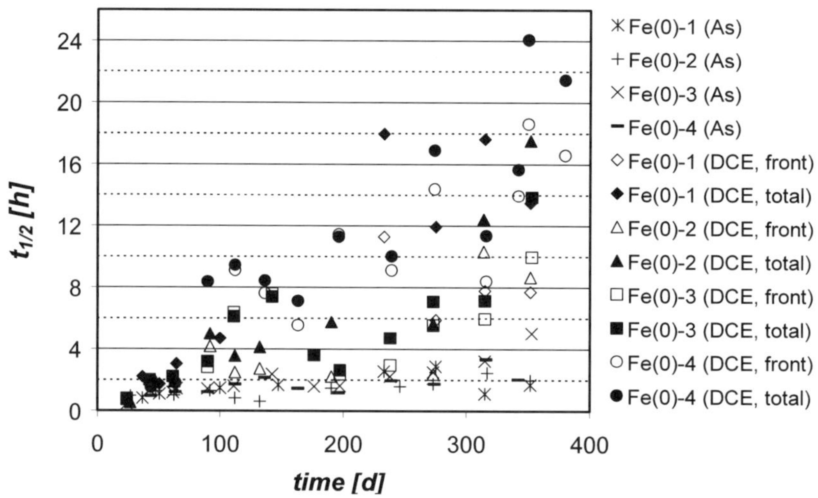

Fig. 6 Half-life times for arsenic removal and 1,2-DCE(cis) degradation in Fe^0-1 to Fe^0-4. DCE half-life times were calculated from the input until degradation terminates or concentrations reach the detection limit (total), and only for the flow path where no or minor inhibition is presumed (front).

to the more complex conditions of our experiments, including diverse kinds of precipitation, As removal could not be differentiated into zero order and first order dominated areas. In most cases, simple first order kinetics described As removal. Independent of the input solution chemistry, half-life times for As removal increased in the beginning and stabilized after approximately 150 days at an average value of 2 h (Fig. 6).

1,2-DCE(*cis*) was dechlorinated by Fe^0 in all Fe^0 filled columns despite the concurrent sorption of As (shown for Fe^0-1 in Fig. 7). Dechlorination was proven by the temporary production of vinyl chloride which was in turn degraded to below the detection limit within the first half of the flow path. DCE degradation in Fe^0-1 was stable during the first two months and restricted to the first 20 cm of the column's length. After 100 days DCE degradation was increasingly inhibited. However, a zone where DCE was still degraded moved in the direction of water flow and possibly stabilized at approximately 300 days. Si adsorption showed comparable behaviour with stable removal until 65 days within the first centimetres followed by a migration of the Si removal and a stabilization of the zone where Si finally adsorbs. Si is known as a strong corrosion inhibitor and Si was shown to decrease the degradation of VOC by Fe^0 (e.g. Kohn *et al.*, 2003). Since DCE degradation was strongly inhibited in areas where Si was adsorbed in our experiments, it is assumed that spreading occupancy of surfaces by Si is the reason for the migrating and decreasing DCE degradation. Other explanations for this inhibition, such as carbonate or sulphide precipitation, are less plausible but cannot be excluded. The reason for the migration of Si removal in Fe^0-1 is thought to be the onset of MSR that started between 65 and 100 days and corresponds with the time span when DCE and Si removal began to migrate. The reduction of sulphate to sulphide can decrease the corrosion induced iron (hydr-)oxide formation because fast precipitation of iron sulphides reduces available Fe^{2+} and Fe^{3+}. Iron (hydr-)oxide precipitation that initially produced enough new sorption sites for Si adsorption within the first centimetres obviously decreased in this area with the onset of MSR. Such a change can cause the migration of the Si adsorption into the area where sulphide formation was largely completed and where apparently enough new iron (hydr-)oxides were generated. Since sulphate is present in most groundwaters, MSR generally proceeds in Fe^0-PRBs and will be an important controlling factor for the long-term degradation of VOC by Fe^0.

In contrast to Si, As is still removed within the first centimetres despite the presumed iron sulphide precipitation and decreased iron (hydr-)oxide formation. One reason for this is precipitation of arsenic sulphides that had been confirmed by XANES spectra received from XAS at samples from areas where sulphide is present (Fig. 8). Preliminary examination of EXAFS spectra from these samples gave evidence that As is additionally associated with iron sulphides. Further analysis of EXAFS spectra and further solid phase investigations to reinforce and refine these findings are under way.

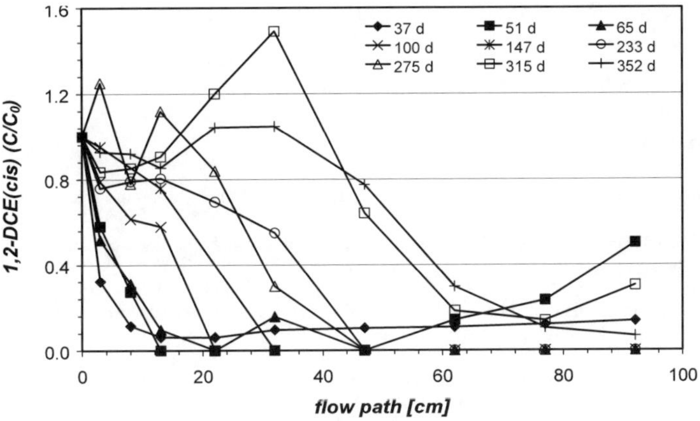

Fig. 7 1,2-DCE(cis) concentrations in Fe^0-1 normalized to the input concentration.

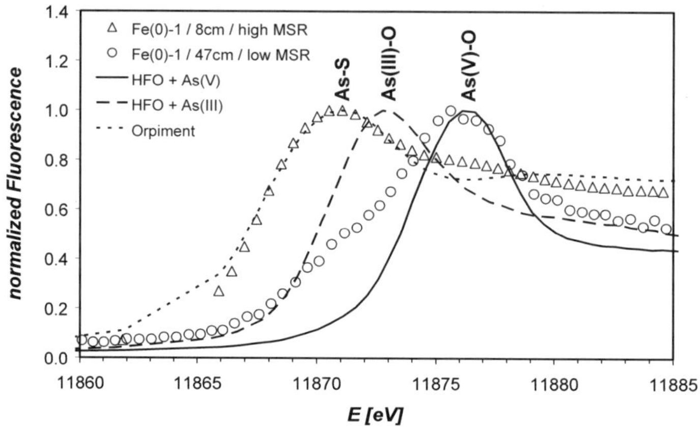

Fig. 8 XANES spectra of standard substances and two samples from Fe^0-1 which were taken in distances of 8 and 47 cm from the input.

Our findings indicate that outflow As concentrations of 1–10 μg l^{-1} can be maintained for years or maybe even decades, if retention times are appropriately chosen to allow establishment of equilibrium between dissolved As and the solid phase. No signs of breakthrough or reactivity loss had been observed during a period of more than 400 days, which is presumably a consequence of the formation of arsenic sulphides. Should the reactivity of Fe^0 become exhausted, Si can be used as natural reactive tracer for the removal of As as indicated by Fig. 3. Breakthrough of input Si concentrations does not necessarily signify the impending As breakthrough, but can be regarded as an early-warning. On the other hand, low outflow Si concentrations are an indication of sufficient remaining reactivity of Fe^0 for the removal of As.

Acknowledgements The presented work is part of a project (02WR0370) which is funded by the German Federal Ministry for Education and Research (BMBF) and the Industry Waste Society of Hesse (HIM). The project is part of the German PRB network, RUBIN. The authors gratefully acknowledge J. Mattusch and B. Daus from the Environmental Research Centre Halle/Leipzig GmbH (UFZ) for arsenic species measurements.

REFERENCES

Dixit, S. & Hering, J. J. (2003) Comparison of arsenic (V) and arsenic(III) sorption onto iron oxide minerals: implications for arsenic mobility. *Environ. Sci. Technol.* **37**(18), 4182–4189.

Farrell, J., Wang, J., O'Day, P. & Conklin, M. (2001) Electrochemical and spectroscopic study of arsenate removal from water using zero-valent iron media. *Environ. Sci. Technol.* **35**(10), 2026–2032.

Khan, A. H., Rasul, S. B., Munir, A. K. M., Habibuddowla, M., Alauddin, M., Newaz, S. S. & Hussam, A. (2000) Appraisal of a simple arsenic removal method for groundwater of Bangladesh. *J. Environ. Sci. Health* **35**(7), 1021–1042.

Klausen, J., Vikesland, P. J., Kohn, T., Burris, D. R., Ball, W. P. & Roberts, A. L. (2003) Longevity of granular iron in groundwater treatment processes: solution composition effects on reduction of organohalides and nitroaromatic compounds. *Environ. Sci. Technol.* **37**(6), 1208–1218.

Köber, R., Ebert, M., Schlicker, O. & Dahmke, A. (2002) Degradation of chlorinated ethenes by Fe(0)–inhibition processes and mineral precipitation. *Environ. Geol.* **41**(6), 644–452.

Köber, R., Giarolli, F. & Dahmke, A. (2003) Developement of up- and downstream permeable reactive barrier systems for remediation of arsenic groundwater contaminations. In: *Sixth International Symposium and Exhibition on Environmental Contamination in Central and Eastern Europe and the Commonwealth of Independent States* (1–4 September, Prague, Czech Republic), Paper 260.

Köber, R., Giarolli, F. & Dahmke, A. (2004) Up- and downstream PRBs for remediation of arsenic and VOC. In: *Remediation of Chlorinated and Recalcitrant Compounds* (ed. by A. R. Gavaskar & A. S. C. Chen). (Fourth Int. Conf., 24–27 May, Monterey, Canada), Paper 2A-22.

Kohn, T., Kane, S. R., Fairbrother, D. H. & Roberts, A. L. (2003) Investigation of the inhibitory effect of silica on the degradation of 1,1,1-trichloroethane by granular iron. *Environ. Sci. Technol.* **37**(24), 5806–5812.

Lackovic, J. A., Nikolaidis, N. P. & Dobbs, G. M. (2000) Inorganic arsenic removal by zero-valent iron. *Environ. Engng Sci.* **17**(1), 29–39.

Manning, B. A., Hunt, M. L., Amrhein, C. & Yarmoff, J. A. (2002) Arsenic(III) and arsenic(V) reactions with zero valent iron corrosion products. *Environ. Sci. Technol.* **36**(24), 5455–5461.

Melitas, N., Conklin. M. & Farrell, J. (2002a) Electochemical study of arsenate and water reduction of iron media used for arsenic removal from potable water. *Environ. Sci. Technol.* **36**(14), 3188–3193.

Melitas, N., Wang, J., Conklin, M., O'Day, P., & Farrell, J. (2002b) Understanding soluble arsenate removal kinetics by zerovalent iron media. *Environ. Sci. Technol.* **36**(9), 2074–2081.

Munir, A. K. M., Rasul, S. B., Habibuddowla, M., Alauddin, M., Hussam, A., & Khan, A. H. (2001) Evaluation of performance of sono 3-kolshi filter for arsenic removal from groundwater using zero valent iron through laboratory and field studies. In: *Technologies for Arsenic Removal From Drinking Water* (Dhaka, Bangladesh, 5–7 May). Bangladesh University of Engineering and Technology and The United Nations University of Tokyo.

Nikolaidis, N. P., Dobbs, G. M. & Lackovic, J. A. (2003) Arsenic removal by zero-valent iron: field, laboratory and modeling studies. *Water Res.* **37**, 1417–1425.

Raven, K. P., Jain, A. & Loeppert, R. H. (1998) Arsenite and arsenate adsorption on ferrihydrite: kinetics, equilibrium, and adsorption envelopes. *Environ. Sci. Technol.* **32**(3), 344–349.

Su, C. & Puls, R. W. (2001a) Arsenate and arsenite removal by zerovalent iron: effects of phosphate, silicate, carbonate, borate, sulfate, chromate, molybdate, and nitrate, relative to chloride. *Environ. Sci. Technol.* **35**(22), 4562–4568.

Su, C. & Puls, R. W. (2001b) Arsenate and arsenite removal by zerovalent iron: kinetics, redox transformation, and implications for *in situ* groundwater remediation. *Environ. Sci. Technol.* **35**(7), 1487–1492.

Su, C. & Puls, R. W. (2003) *In situ* remediation of arsenic in simulated groundwater using zerovalent iron: laboratory column tests in combined effects of phosphate and silicate. *Environ. Sci. Technol.* **37**(11), 2582–2587.

Vlassopoulos, D., Pochatila, J., Lundquist, A., Andrews, C. B., Rafferty, M. T., Chiang, K., Sorel, D. & Nikolaidis, N. (2002) An elemental iron reactor for arsenic removal from groundwater. In: *Remediation of Chlorinated and Recalcitrant Compounds* (Monterey, Canada, 20–23 May). Battelle Press, Ohio, USA.

Permeable Reactive Barriers (Proceedings of the International Symposium held at Belfast, Northern Ireland, March 2004). IAHS Publ. 298, 2005.

Remediation of chlorinated aliphatic hydrocarbons in groundwater using Fe^0 PRB

KEITH C. K. LAI[1], IRENE M. C. LO[1] & PETER KJELDSEN[2]

1 *Department of Civil Engineering, Hong Kong University of Science and Technology, Clear Water Bay, Kowloon, Hong Kong*
cemclo@ust.hk

2 *Environment & Resources DTU, Technical University of Denmark, Lyngby, Denmark*

Abstract A funnel-and-gate permeable reactive barrier (PRB) packed with zero valent iron (Fe^0) was constructed at Vapokon site, Denmark in September 1999 to remediate the groundwater contaminated by chlorinated aliphatic hydrocarbons (CAHs). In order to understand the variation of the performance of the PRB over the past 4 years and identify the major factor(s) affecting the iron barrier's performance, comprehensive monitoring of the behaviour of the Fe^0 PRB was conducted between March 2000 and August 2003. Although there was a continuous decrease in total alkalinity (90.3%), calcium (81.7%) and sulphate (69.2%) ions in the groundwater across the PRB, probably caused by mineral precipitation and resulting in 0.88% porosity loss per year, no noticeable trend of deterioration of the barrier's performance was observed. Virtually, over the past 4 years, about 92.4–97.5% removal of CAHs from the groundwater could be achieved across the PRB, except in September 2002 and January 2003. Instead, climatic variation of the barrier's performance on CAHs dechlorination was examined in which the dechlorination rates in cold season (January 2003 and March 2000) were generally smaller than those in hot season (August 2003, September 2000 and September 2001). The high DO content in the groundwater caused by the recharge from rainfall precipitation may also be one of the factors affecting the performance of PRB on CAHs dechlorination.

Key words chlorinated aliphatic hydrocarbons; climatic variation; funnel-and-gate; Vapokon site; zero valent iron

INTRODUCTION

Chlorinated aliphatic hydrocarbons (CAHs) such as tetrachloroethene (PCE), trichloroethene (TCE) and 1,1,1-trichloroethane (1,1,1-TCA) are the most common contaminants found in groundwater because of their wide applications as cleaning and degreasing agents (O'Hannesin & Gillham, 1998). Since most of them are carcinogenic and also highly recalcitrant in subsurface environments, their drinking water limits are generally in an order of a few micrograms per litre (Ramamoorthy & Ramamoorthy 1997).

The increase of popularity of permeable reactive barrier (PRB) technology packed with zero valent iron (Fe^0) on the remediation of CAHs polluted groundwater is stemmed from its advantages over conventional pump-and-treat systems including the absence of invasive surface structures and equipment as well as low operation and maintenance costs (EnviroMetal Inc., 1998). The oxidation of Fe^0 by the groundwater results in chemically reducing conditions and electron source for the reduction of CAHs into innocuous form with concomitant release of hydroxide ion and hydrogen gas (USEPA 1998; Yabusaki *et al.* 2001). Hydrogenolysis (equation (1)), reductive-β-elimination (equation (2)) and direct catalytic hydrogenation by hydrogen gas (equation (3)) are the three main pathways involved in the dechlorination reactions (Matheson & Tratnyek, 1994; USEPA 1998). However, the problem of mineral precipitation on Fe^0 surfaces under high pH and low redox potential (Eh) conditions can potentially passivate the Fe^0, thereby adversely influencing the barrier's long-term performance (Mackenzie *et al.*, 1999). In addition, climatic variation of groundwater characteristics such as the high dissolve oxygen (DO) content in groundwater during the rainy season and the low groundwater temperature in winter is able to cause fluctuation in a barrier's performance (Helland *et al.*, 1995; Su & Pul, 1999).

$$\mathrm{RHCl=RHCl+2e^- + H^+ \rightarrow RH_2 = RHCl + Cl^-} \quad (1)$$

$$\mathrm{RHCl=RHCl+2e^- \rightarrow RH \equiv RH + 2Cl^-} \quad (2)$$

$$\mathrm{RHCl=RHCl+H_2 \xrightarrow{Catalyst} RH_2 = RHCl + H^+ + Cl^-} \quad (3)$$

Performance assessment of a funnel-and-gate Fe^0 PRB installed at the Vapokon site of Denmark was conducted on six occasions between March 2000 and August 2003 so as to examine the variation of the performance of PRB on dechlorinating CAHs. Identification of the major factor(s) affecting the barrier's performance and investigation of the geochemical variation of the groundwater along the PRB are also the objectives of this field study.

FIELD DESCRIPTION

The Vapokon site is located in the northern part of Fuen Island, Denmark. Its geological environment mainly consists of a layer of loam and soil filings extending approx. 1.5 m from the ground surface, which has an altitude of about 22.22 m a.s.l. Just below the vadose zone, there is a water saturated sandy aquifer, approx. 10 m in thickness, underlain by a relatively impermeable clay layer with a minimum thickness of about 15 m. The groundwater flows in a southeastern direction at a seepage velocity of 400–500 m year^{-1} towards a rain water pond nearby (Lai *et al.*, 2004). Because of the spill from an open underground storage tank, there was heavy soil and groundwater contamination with CAHs and aromatic organics (BTEX), resulting in a 80 × 200 m contaminant plume (Fig. 1(a)). As shown in Table 1, PCE, TCE, 1,1,1-TCA, *cis*-dichloroethene (cis-DCE) and dichloromethane (DCM) were observed to be the major CAHs found in the contaminant plume accompanying trace amounts of vinyl chloride (VC), 1,1-dichloroethene (1,1-DCE), trans-dichloroethene (trans-DCE), 1,1-dichloroethane (1,1-DCA), 1,2-dichloroethane (1,2-DCA) and chloroform (TCM). For the sake of remedying this contaminated site, 4600 t of contaminated soils were first excavated and a Fe^0 PRB was then installed to remediate CAHs contaminated groundwater.

As can be seen from Fig. 1(b), the funnel-and-gate Fe^0 PRB, which was installed at Vapokon site in September 1999, is composed of two 110–130 m long sheet pilings (funnel) and a reactive gate containing 270 t of Fe^0 (ETI-CC-1004, Connelly-GPM Inc.). The reactive gate, which was 14.5 m wide, 9 m deep and 0.8 m thick possessed an average hydraulic conductivity of 20.28 m day^{-1}, a porosity of 0.598 as well as ratio of Fe^0 surface area to solution volume (λ_s) of 7.79 m^2 ml^{-1}. Gravel packs with 1 m

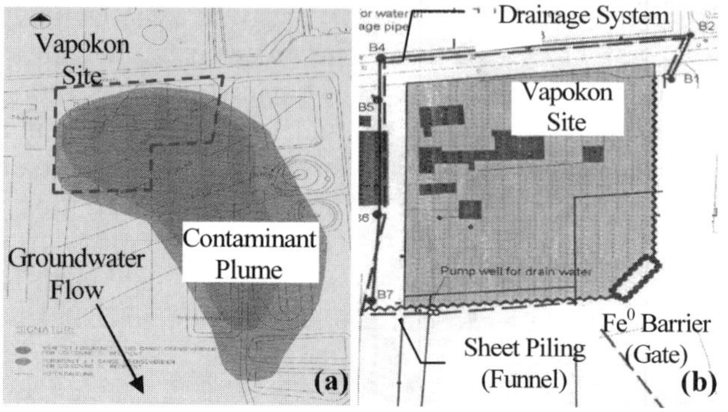

Fig. 1 (a) Area of the contaminant plume, and (b) the funnel-and-gate Fe^0 PRB at Vapokon site.

Parameters/date	March 2000	September 2002	August 2003
PCE	12.00	4.81	7.30
TCE	5.00	23.66	22.00
cis-DCE	5.50	29.54	38.00
VC	0.55	na	2.40
1,1-DCE	0.04	na	0.61
trans-DCE	0.10	nd	0.43
1,1,1-TCA	4.40	15.47	31.00
1,1-DCA	0.15	0.70	0.66
1,2-DCA	6.10	nd	0.03
TCM	0.64	nd	0.26
DCM	2.90	1.69	1.30

Note: The units of the CAHs concentrations are in mg l^{-1}. na, not available; nd, not detectable.

thickness were placed just upgradient and downgradient of the reactive barrier so as to spread the contaminated groundwater uniformly across the PRB. Besides, in order to prevent any uncontaminated groundwater from flowing into the capture zone of the funnel-and-gate system, a groundwater drainage system was installed in the upgradient of the Fe^0 PRB. Based upon the results from the natural gradient tracer study for the PRB at Vapokon site (Lai *et al.*, 2004), a groundwater seepage velocity of about 99.5 m $year^{-1}$ was found within the PRB, which was about 4–5 times smaller than its natural groundwater flow.

METHODS AND MATERIALS

Installation of monitoring network and groundwater sampling

Two types of monitoring network were installed across the Fe^0 PRB, resulting in six sampling fences located 0, 0.1, 0.4, 0.58, 0.7 and 1.1 to 5.8 m apart from the upgradient edge of the PRB. The 15 monitoring wells constructed at the time of the PRB installation had 45 piezometers, each ending with a 1-m long screen. The three piezometers in each well generally screened at depths of 12.52 to 14.22, 15.62 to 16.47 and 18.02 to 19.12 m a.s.l., respectively. In addition, 20 nylon and 6 copper multilevel samplers with 310 screens distributing in approximately 13.78 to 20.28 m a.s.l. were also installed in the central part of the PRB in September 2002 (Lai *et al.* 2004). Groundwater samplings were conducted in March 2000, September 2000, September 2001, September 2002, January 2003 and August 2003 in order to obtain long-term monitoring data.

Groundwater sample analyses

For the measurement of CAHs in groundwater, gas chromatography (HRGC 5300, Carlo Erba Instruments) equipped with an electron capture detector (ECD-400, Carlo Erba Instruments) and the method of liquid–liquid extraction using pentane (Pestiscan, LAB-SCAN) as solvent were applied. Detection of calcium (Ca^{2+}), magnesium (Mg^{2+}) and total iron (Fe_T) as well as chloride (Cl^-) and sulphate (SO_4^{2-}) ions were implemented by using atomic absorption spectrometer (Perkin Elmer 5000, Perkin Elmer) and ion chromatography (LC20 Chromatography, Dionex), respectively. Besides, titration method using 0.02 N sulphuric acid (UN1830, General Chemical) as titrant and Bromcresol green (B-0443-100, Exaxol Chemical Corporation) as indicator was applied for total alkalinity (TAL) measurement. Physical characteristics of Vapokon groundwater were measured immediately in the field by transferring the groundwater collected from the monitoring wells to a flow cell equipped with pH (pH330 SET-1, WTW), DO (OXI330-SET, WTW), conductivity (LF330-SET, WTW) and Eh (Ag/AgCl 800-323-4340, Cole-Parmer) probes. Readings were taken when three successive readings were within 10% min^{-1} of variation.

DATA ANALYSIS

Regarding the kinetic, Gillham & O'Hannesin (1994) have found out that the dechlorination reaction is a pseudo-first-order reaction with respect to contaminant concentration. Because of sequential dechlorination of major chlorinated organics (MCOs) such as PCE, TCE, 1,1,1-TCA and TCM into the corresponding chlorinated daughter products (CDPs) including cis-DCE, 1,1-DCE, VC, 1,1-DCA and DCM (Vogel & McCarty 1987; Helland *et al.* 1995; Su & Puls 1999), the following sequential dechlorination models (models 1, 2 and 3), which take account of concurrent production and dechlorination of CAHs, were applied for the computation of dechlorination rate constant (k_{obs}). The first-order equations for the models 1, 2, and 3 are illustrated in equations (4) to (8), (10) and (11), and (12), respectively. Since sequential transformations of PCE and TCE into trans-DCE by Fe^0 have not been proven (Su & Puls, 1999), a first-order equation solely considering dechlorination of trans-DCE was applied for determining $k_{trans-DCE}$ (equation (9)). Besides, computation of the k_{obs} of 1,2-DCA and DCM was not considered since they have been substantiated to be not treatable by Fe^0 (USEPA, 1998). With the application of the f_{PCE1} (0.983), f_{PCE2} (0.017), f_{TCE1} (0.014), f_{TCE2} (0.003), $f_{cis\text{-}DCE}$ (0.048) and $f_{1,1,1\text{-}TCA}$ (0.441) previously found by EnviroMetal Technologies Inc. (1998), equations (1)–(9) were used directly in Scientist®, which can perform integration and thereby calculated the k_{obs} by fitting the integrated solutions to the relevant experimental data (MicroMath, 1995).

Model 1

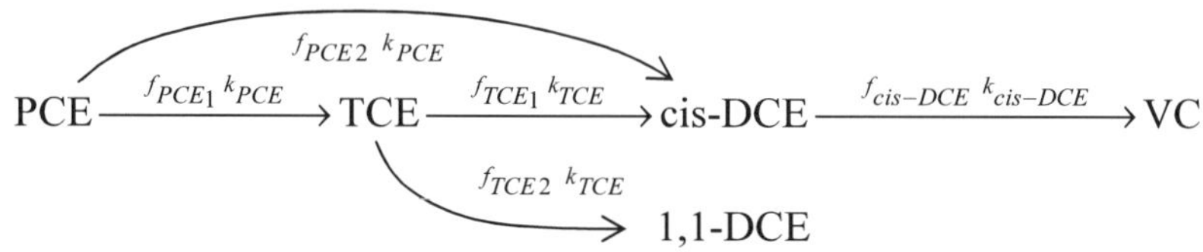

$$\frac{d[PCE]}{dt} = -k_{PCE}[PCE] \tag{4}$$

$$\frac{d[TCE]}{dt} = f_{PCE1}k_{PCE}[PCE] - k_{TCE}[TCE] \tag{5}$$

$$\frac{d[cis-DCE]}{dt} = f_{PCE2}k_{PCE}[PCE] + f_{TCE1}k_{TCE}[TCE] - k_{cis-DCE}[cis-DCE] \tag{6}$$

$$\frac{d[VC]}{dt} = f_{cis-DCE}k_{cis-DCE}[cis-DCE] - k_{VC}[VC] \tag{7}$$

$$\frac{d[1,1-DCE]}{dt} = f_{TCE2}k_{TCE}[TCE] - k_{1,1-DCE}[1,1-DCE] \tag{8}$$

$$\frac{d[trans-DCE]}{dt} = -k_{trans-DCE}[trans-DCE] \tag{9}$$

Model 2 $\quad \text{1,1,1-TCA} \xrightarrow{f_{1,1,1-TCA}\,k_{1,1,1-TCA}} \text{1,1-DCA}$

$$\frac{d[1,1,1-TCA]}{dt} = -k_{1,1,1-TCA}[1,1,1-TCA] \tag{10}$$

$$\frac{d[1,1-DCA]}{dt} = f_{1,1,1-TCA}k_{1,1,1-TCA}[1,1,1-TCA] - k_{1,1-DCA}[1,1-DCA] \tag{11}$$

Model 3 $\quad \text{TCM} \xrightarrow{f_{TCM}\,k_{TCM}} \text{DCM}$

$$\frac{d[TCM]}{dt} = -k_{TCM}[TCM] \tag{12}$$

where f is the molar fraction; k is the k_{obs} of different CAHs; [   ] is the CAHs concentration (mg l^{-1}); t is time.

Since the k_{obs} was found directly proportional to λ_s, the dechlorination rate constant normalized to the ratio of Fe^0 surface area to solution volume (k_{sa}), which can be obtained from equation (13), is usually reported in the literature for comparison.

$$k_{sa} = \frac{k_{obs}}{\lambda_s} \tag{13}$$

Table 2 Variations of inorganic and physical characteristics of Vapokon groundwater along the Fe^0 PRB.

Parameters /Sampling Time	Ca^{2+}	Mg^{2+}	TAL	Fe_T	SO_4^{2-}	Cl^-	pH	Eh (mV)	Spec. Cond. (μS cm^{-1})
March 2000	**510**	**64.9**	**372**	**13.2**	**189**	**50.3**	**7.03**	na	**1153**
	160	37.2	77.7	8.12	72.8	43.1	8.98		463
	146	21.4	19.3	0.41	108	42.0	9.50		432
September 2000	**646**	**52.1**	**323**	**13.2**	**145**	**43.6**	**7.38**	na	**644**
	133	41.2	99.2	9.83	47.6	55.4	9.69		330
	81.0	27.7	30.7	0.60	39.6	47.8	10.56		238
September 2001	na	na	na	na	na	na	**7.33**	na	**679**
							9.53		378
							10.62		230
September 2002	**338**	na	na	**0.18**	**27.2**	**46.1**	na	na	na
	136			0.25	3.70	65.2			
	53.5			0.13	2.20	38.0			
January 2003	**299**	na	**248**	**0.15**	na	**43.9**	**7.16**	**105**	**646**
	95.9		125	0.13		38.9	9.44	−115	309
	48.2		35.4	0.38		33.9	10.21	−150	229

Note: The units of the inorganic parameters are in mg l^{-1}. na, not available. Spec. Cond., specific conductance. The figures shown for each sampling, from the top to bottom, represent the average value of the parameters collected from the upgradient location, middle and downgradient part of the Fe^0 PRB, respectively. The bold figures represent the original characteristics of the groundwater.

RESULTS AND DISCUSSION

Performance of the funnel-and-gate Fe^0 PRB on CAHs dechlorination

Over the past four years, rapid dechlorination of MCOs and CDPs really occurred across the PRB. As shown in Fig. 2(a), in January 2003, the sum of maximum concentration of MCOs in Vapokon groundwater was about 38 mg l^{-1} in the immediate vicinity of the front edge of the Fe^0 PRB. Along the reactive barrier, the MCOs concentration dropped tremendously and it was less than 0.6 mg l^{-1} just downgradient of the Fe^0 PRB. However, because of high background concentration and significant spatial variation in Cl^- concentration in the groundwater fluctuating between 6.32 and 113 mg l^{-1} (Table 2), concurrent increase in Cl^- concentration could not be observed.

For the sake of understanding the variation of the barrier's performance on CAHs dechlorination, the k_{sa} of different CAHs computed between March 2000 and August 2003 were compared. As can be seen from Table 3, over the past four years, no pronounced trend of deterioration of the iron barrier's performance on CAHs dechlorination was observed, even though there was continuous diminution in dissolved ions in the groundwater along the barrier probably caused by mineral precipitation (Table 2). Instead, fluctuation of the Fe^0 reactivity on CAHs dechlorination was examined, particularly for PCE, TCE and 1,1,1-TCA in which their k_{sa} obtained in September 2000, September 2001 and August 2003 were about 3- to 4-fold higher than those in March 2000, September 2002 and January 2003, respectively. The k_{sa} of cis-DCE and trans-DCE, on the contrary, were fairly stable with standard deviations of about 0.0005 and 0.0029 ml m^{-2} h^{-1}, respectively.

Despite the fact that low Fe^0 reactivity on PCE, TCE and 1,1,1-TCA dechlorination was recorded in March 2000, September 2002 and January 2003, the removal efficiency of MCOs across the reactive barrier did not suffer severe deterioration at

Table 3 Variations of k_{sa} ($\times 10^{-2}$) of different CAHs between March 2000 and August 2003.

Parameters/Date		March 2000	September 2000	September 2001	September 2002	January 2003	August 2003
PCE	k_{sa}	3.90	7.59	10.00	1.34	1.39	5.04
	R^2	1.000	0.987	0.989	0.955	0.993	1.000
TCE	k_{sa}	3.64	8.01	10.09	2.36	1.33	4.36
	R^2	0.999	0.984	0.940	0.992	0.998	1.000
cis-DCE	k_{sa}	0.18	0.17	0.21	0.17	0.14	0.28
	R^2	0.930	0.861	0.569	0.835	0.851	0.632
VC	k_{sa}	0.95	0.60	0.34	na	na	0.64
	R^2	0.824	0.741	0.766	na	na	0.933
1,1-DCE	k_{sa}	0.43	2.28	5.98	na	na	1.66
	R^2	0.850	0.965	0.915	na	na	0.975
trans-DCE	k_{sa}	5.38	nc	5.11	nc	nc	4.80
	R^2	0.999	nc	0.760	nc	nc	0.998
1,1,1-TCA	k_{sa}	1.67	6.89	6.30	2.14	1.12	7.43
	R^2	1.000	0.996	0.987	0.694	0.998	0.999
1,1-DCA	k_{sa}	0.85	2.62	nc	nc	0.19	3.36
	R^2	0.975	0.951	nc	nc	0.524	0.786
TCM	k_{sa}	0.91	2.55	4.67	nc	nc	1.71
	R^2	0.993	0.944	0.645	nc	nc	0.899

Note: na, not available. nc, not calculable. The unit of k_{sa} is ml m^{-2} h^{-1}. R^2 means coefficient of variation. The k_{sa} for every CAHs was calculated from their maximum concentration profiles along the Fe^0 PRB.

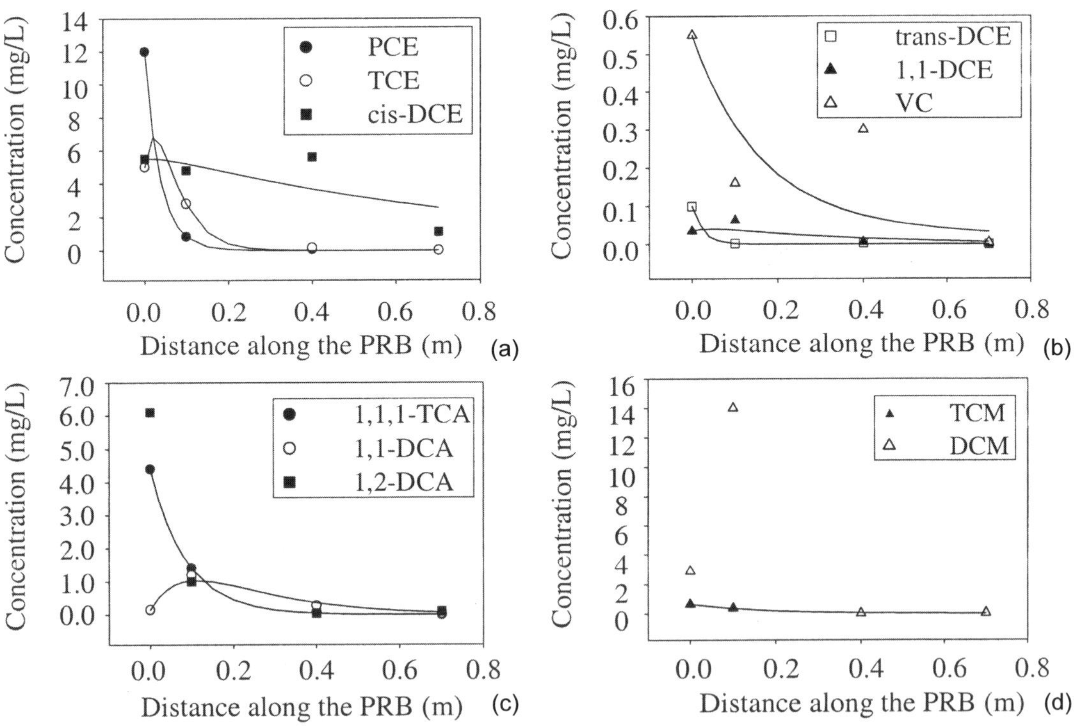

Fig. 3 Maximum concentration of (a) PCE, TCE and cis-DCE, (b) trans-DCE, 1,1-DCE and VC, (c) 1,1,1-TCA, 1,1-DCA and 1,2-DCA as well as (d) TCM and DCM along the Fe^0 PRB in March 2000. The lines represent the best fit of the sequential dechlorination model to the data.

those times. On average, between March 2000 and August 2003, about 97.5% of MCOs in groundwater could be removed across the PRB. The insignificance of the influence of the low Fe^0 reactivity on MCOs removals is most likely due to that according to Fig. 3(a)–(d), most of MCOs were observed to be dechlorinated into non-detectable levels in the first half of the reactive barrier. Therefore, even though there was deterioration in the rate of PCE, TCE and 1,1,1-TCA dechlorination in March 2000, September 2002 and January 2003, the downgradient part of the reactive barrier could still provide enough residence time for the effective removal of MCOs, thereby maintaining high efficiency on MCOs removal throughout the whole study period.

Effective removal of CDPs could also be achieved across the PRB in March 2000, September 2000, September 2001 and August 2003 in which approximately 92.4% of CDPs in Vapokon groundwater was dechlorinated into benign products. Because of the sequential dechlorination from the corresponding MCOs, there were intermediate increases in CDPs concentrations in the upgradient part of the PRB prior to the significant drop of the concentrations in the downgradient part of the reactive barrier (see Fig. 3(a)–(d)). However, in September 2002 and January 2003, poor performance of the Fe^0 PRB on CDPs dechlorination was observed. As shown in Fig. 2(b), in January 2003, the peak concentration of sum of CDPs within the PRB was examined appearing at the end of the PRB in lieu of the upgradient part of the reactive barrier. Similar phenomenon could also be observed in September 2002.

As can be seen from Table 1, *cis*-DCE was the dominant CDP in the contaminant plume and its rate of dechlorination between March 2000 and August 2003, as aforementioned, was quite stable (Table 3). Therefore, the appearing of the peak concentration of CDPs or cis-DCE at the end of the PRB in September 2002 and January 2003 was not due to the low Fe^0 reactivity on cis-DCE dechlorination. In fact, the low rates of dechlorination of PCE and TCE in September 2002 and January 2003 are believed to be one of the underlying reasons causing the poor performance of the PRB on CDPs removals since PCE and TCE were dechlorinated into *cis*-DCE within the Fe^0 PRB (EnviroMetal Technologies Inc., 1998). Therefore, the reduction of PCE and TCE k_{sa} in September 2002 and January 2003 consequently postponed the appearing of the peak concentration of *cis*-DCE from the upgradient part to the end of the reactive barrier, resulting in the increase in *cis*-DCE concentration across the PRB.

However, the low Fe^0 reactivity on PCE and TCE dechlorination in March 2000 did not adversely influence the removal efficiency of CDPs at that moment in which 92.8% or thereabouts of CDPs could still be removed from Vapokon groundwater across the barrier. The insignificance of the influence from the low k_{sa} of PCE and TCE on the CDPs removals was believed to pertain to the low CAHs concentrations in Vapokon groundwater in March 2000 (Table 1). As alluded to above, 4600 t of heavily contaminated soil were first excavated from the hotspot prior to the installation of the Fe^0 PRB in September 1999. Most likely owing to the removal of the main source of contamination and slow rate of desorption of CAHs from the remaining contaminated soil, the CAHs concentration in Vapokon groundwater was relatively low in March 2000 compared with those between September 2000 and August 2003. Low PCE and TCE concentrations in groundwater consequently led to minor production of *cis*-DCE within the reactive barrier, thereby indicating an insignificant contribution from PCE and TCE dechlorination on the appearance of the peak concentration of *cis*-DCE

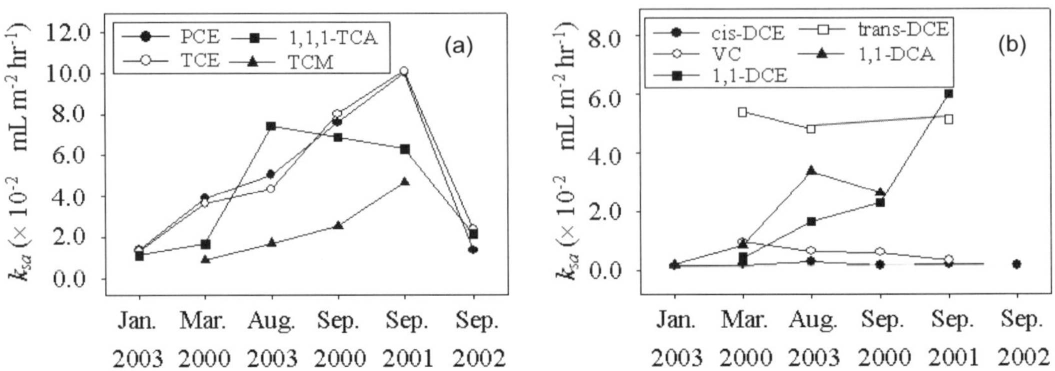

Fig. 4 Climatic variation of the k_{sa} of (a) PCE, TCE, 1,1,1-TCA and TCM and (b) cis-DCE, VC, 1,1-DCE, trans-DCE and 1,1-DCA.

within the PRB. As a consequence, no significant detrimental effect on the CDPs removals was exerted from the low k_{sa} of PCE and TCE in March 2000.

Climatic variation of the performance of the Fe^0 PRB

Careful examination of Table 3 indicates that, except for the results in September 2002, the k_{sa} of different CAHs obtained in the cold season (i.e. January 2003 and March 2000) were usually smaller than those received in the hot season (i.e. August 2003, September 2000 and September 2001). As you can see from Fig. 4(a) and (b), excluding the k_{sa} obtained in September 2002, the k_{sa} generally increased in the following order: January, March, August, September, thereby manifesting climatic variation of the barrier's performance on CAHs dechlorination. Probably owing to the low temperature in groundwater in cold season, the Fe^0 reactivity on CAHs dechlorination was comparatively low in January 2003 and March 2000, since the dechlorination reactions of CAHs by Fe^0 have been proven primarily to be reaction-limited kinetic (Su & Puls, 1999). Low temperature in groundwater consequently lowered the rate of oxidation of Fe^0 or release of electron from Fe^0 and thereby adversely affected the rate of dechlorination reactions in cold season.

However, the abnormal low rate of dechlorination in September 2002 was most likely attributed to the high DO content in the groundwater at that moment. During the day of groundwater sampling in September 2002, because of the recharge from rainfall precipitation, the DO content of the groundwater, on average, was about 6.25 mg l^{-1} (4.50–9.90 mg l^{-1}), which was much higher than the usual DO content measured in Vapokon groundwater (i.e. about 0.53 to 1.53 mg l^{-1}). The DO in the groundwater did not just compete with CAHs for the electron released from Fe^0; it could reduce the availability of the reactive site on the Fe^0 surface for the dechlorination reaction and inhibit the binding rate between Fe^0 surface and CAHs, thereby leading slower dechlorination kinetic in September 2002 (Helland *et al.*, 1995).

Geochemical variation of Vapokon groundwater along the PRB

The pH conditions of between 9.50 and 10.62, as well as about –150 mV of Eh within the PRB, highly encouraged mineral precipitation (Yabusaki *et al.*, 2001). As a consequence, across the PRB, there were approx. 81.7% decrease in Ca^{2+}, 57.0% reduction in Mg^{2+}, 90.3% diminution in TAL as well as 69.2% drop of SO_4^{2-}, resulting in 64.1% decrease in specific conductance in the groundwater (Table 2). As Ca^{2+}, CO_3^{2-} and SO_4^{2-} were the major dissolved ions to be precipitated out across the PRB, it is believed that calcite ($CaCO_3$) and pyrite (FeS) were the dominated precipitates formed inside the PRB. By assuming calcite and pyrite as the only precipitates formed inside the barrier, it was computed that mineral precipitation, in total, caused 2.14 m^3 loss of void volume throughout the reactive barrier between March 2000 and August 2003, which corresponded to 3.46% total loss of porosity or 0.88% loss per year. Except for carbonate-containing green rust and magnetite (Fe_3O_4), all the mineral precipitates possibly formed on the Fe^0 surface have an ability to passivate the Fe^0 since they can prevent the electron transfer between the Fe^0 surface and the contaminants (USEPA, 1998; Ritter *et al.*, 2002). However, the absence of deterioration of the rate of CAHs dechlorination by the Fe^0 over the past four years suggested the insignificance of the passivated effect exerted from the mineral precipitation.

CONCLUSIONS

Effective removal of CAHs could be achieved across the PRB over the past four years; approximately 97.5% of MCOs were removed from the groundwater across the PRB and about 92.4% for CDPs except in September 2002 and January 2003. Although there was a continuous decrease in dissolved ions in the groundwater along the barrier probably caused by mineral precipitation, there was no evident trend of deterioration of the iron barrier's performance on CAHs dechlorination between March 2000 and August 2003, thereby suggesting the insignificance of the passivated effects exerted from mineral precipitation. Instead, over the past four years, climatic variation of the performance of the PRB was examined. In general, most likely owing to the low groundwater temperature in cold season, the k_{sa} of CAHs in January 2003 and March 2000 (cold season) were usually smaller than those in August 2003, September 2000 and September 2001 (i.e. hot season) especially for PCE, TCE and 1,1,1-TCA. The unusual low rate of dechlorination in September 2002 was probably due to the high DO content in the groundwater caused by the recharge from rainfall precipitation. Among various mineral precipitates possibly formed on the Fe^0 surface, $CaCO_3$ and FeS are believed to be the dominant precipitates formed within the barrier, which in total caused 3.46% loss of porosity or 0.88% loss per year.

REFERENCES

EnviroMetal Technologies Inc. (1998) *Bench-Scale Treatability Report of the EnviroMetal Process at the Vapokon Site, Fyn, Denmark*, ETI 31438.10, Guelph, Ontario, Canada N1K1S6.

Gillham, R .W. & O'Hannesin, S. F. (1994) Enhanced degradation of halogenated aliphatics by zero valent iron. *Ground Water* **32**(6), 958–967.

Helland, B. R., Alvarez, P. J. J. & Schnoor, J. L. (1995) Reductive dechlorination of carbon tetrachloride with elemental iron. *J. Hazard. Mater.* **41**, 205–216.

Lai, C. K., Lo, I. M. C. & Kjeldsen, P. (2004) Investigation of flow pattern of groundwater through a Fe^0 PRB. In: *Remediation of Chlorinated and Recalcitrant Compounds* (Summary fourth Int. Conf., Monterey, California, USA).

Mackenzie, P. D., Horney, D. P. & Sivavec, T. M. (1999) Mineral precipitation and porosity losses in granular iron columns. *J. Hazard. Mater.* **68**, 1–17.

Matheson, L. J. & Tratnyek, P. G. (1994) Reductive dehalogenation of chlorinated methanes by iron metal. *Environ. Sci. Technol.* **28**(12), 2045–2053.

MicroMath Scientist® (1995) *Experimental Data Fitting/Microsoft Windows Version 2.0*. Salt Lake City, Utah, USA.

O'Hannesin, S. F. & Gillham, R. W. (1998) Long-term performance of an *in situ* "iron wall" for remediation of VOCs. *Ground Water* **36**(1), 164–170.

Ramamoorthy, S. & Ramamoorthy, S. (1997) *Chlorinated Organic Compounds in the Environment—Regulatory and Monitoring Assessment*. Lewis Publishers, Boca Raton, Florida, USA.

Ritter, K., Odziemkowski, M. S. & Gillham, R. W. (2002) An *in situ* study of the role of surface films on granular iron in the permeable iron wall technology. *J. Contam. Hydrol.* **55**, 87–111.

Su, C. & Puls, R. W. (1999) Kinetics of trichloroethene reduction by zerovalent iron and tin: pretreatment effect, apparent activation energy, and intermediate products. *Environ. Sci. Technol.* **33**(1), 163–168.

United States Environmental Protection Agency, Office of Research and Development (1998) Permeable Reactive Barrier Technologies for Contaminant Remediation. *EPA/600/R-98/125, US EPA, Washington, DC 20460, USA.*

Vogel, T. M. & McCarty, P. L. (1987) Abiotic and biotic transformations of 1,1,1-trichloroethane under methanogenic conditions. *Environ. Sci. Technol.* **21**(12), 1208–1213.

Yabusaki, S., Cantrell, K., Sass, B. & Steefel, C. (2001) Multicomponent reactive transport in an *in situ* zero valent iron cell. *Environ. Sci. Technol.* **35**(7), 1493–1503.

Permeable Reactive Barriers (Proceedings of the International Symposium held at Belfast, Northern Ireland, March 2004). IAHS Publ. 298, 2005.

Impact of fouling on the long-term hydraulic behaviour of permeable reactive barriers

LIN LI & CRAIG H. BENSON

Department of Civil and Environmental Engineering, University of Wisconsin-Madison, Madison, Wisconsin 53706, USA

benson@engr.wisc.edu

Abstract This paper describes reactive transport simulations conducted to assess the impact of mineral fouling on the long-term performance of permeable reactive barriers employing granular zero valent iron (ZVI). Three minerals were assumed to form in the pore space ($CaCO_3$, $FeCO_3$, and $Fe(OH)_2$) and the inflowing groundwater was assumed to have the following composition: DO = 10^{-8} M, $Fe^{2+} = 10^{-10}$ M, $Ca^{2+} = 10^{-3}$ M, $OH^- = 10^{-7}$ M, $HCO_3^- = 10^{-3}$ M, and $CO_3^{2-} = 10^{-7}$ M. Results of the simulations show that the porosity and hydraulic conductivity of the ZVI decrease over time and that flows are redistributed throughout the PRB in response to fouling of the pore space. Seepage velocities in the PRB increase, and residence times decrease, due to porosity reductions caused by accumulation of minerals in the pore space. Under the assumed conditions, only subtle changes occur within the first 10 years (i.e. the duration of the current field experience record with PRBs) and the most significant changes do not occur until the PRB has operated for at least 30 years. However, after 30–50 years, reductions in residence time of the order of 50% occurred. More rapid and extensive changes are likely to occur for conditions that result in greater precipitation rates (e.g. groundwater with higher ionic strength, higher velocity).

Key words fouling; iron corrosion; long-term performance; mineral precipitation; permeable reactive barrier; reactive transport modelling; reactive transport; zero valent iron

INTRODUCTION

The permeable reactive barrier (PRB) is a passive treatment technology used to treat contaminated groundwater. PRBs generally are used for long-term treatment and are expected to be in service for decades. During this period, fouling caused by mineral precipitation is a major concern. Fouling causes loss of pore space and reactive surface area of the medium in PRBs, which can alter flow paths, residence times, and the treatment effectiveness of a PRB (Wilkin & Puls, 2003).

Particles of zero valent iron (ZVI) are used as the reactive medium in most PRBs constructed to date. As groundwater flows through the PRB, dissolved oxygen (DO) and water corrode the ZVI, which elevates the groundwater pH and causes precipitation of secondary minerals from the dissolved iron and major ions originally in the inflowing groundwater. Minerals that generally have been identified in PRBs include iron oxides, iron (oxy)hydroxides, and carbonates (Mackenzie *et al.*, 1999; Phillips *et al.*, 2003).

Precipitation of these secondary minerals causes porosity reductions ranging from 0.0007 to 0.03 year^{-1} (Li, 2004). Porosity reductions typically are largest near the entrance of a PRB and diminish with distance into the PRB. The porosity reductions also exhibit spatial variability due to the influence of heterogeneities in aquifer hydraulic properties on flow and transport (Elder *et al*., 2002; Li, 2004).

Although mineral precipitates and porosity reductions have been observed in the field, their effect on the hydraulic behaviour and treatment effectiveness of PRBs over the long term is unknown due to the relatively short history of PRB technology (~12 years in 2004). One approach to long-term assessment is to simulate flow and reactive transport in PRBs using numerical models. In this paper, a modelling study is described that was conducted to evaluate the impact of fouling on hydraulic performance of PRBs over decades of continuous flow in an aquifer having a typical level of heterogeneity.

FLOW AND TRANSPORT MODELLING

The conceptual model is a continuous trench PRB containing ZVI that is placed in a heterogeneous aquifer. Groundwater flowing into the PRB is assumed to be in chemical equilibrium. Major constituents in the inflowing groundwater are assumed to be Fe^{2+}, Ca^{2+}, CO_3^{2-}, OH^-, HCO_3^-, and DO. These constituents are assumed to form three minerals: $CaCO_3$, $FeCO_3$, and $Fe(OH)_2$. These constituents and minerals were selected based on the results of a sensitivity analysis conducted by Li (2004) that included carbonate and sulphide minerals, as well as hydroxides. This analysis showed that $CaCO_3$, $FeCO_3$, and $Fe(OH)_2$ generally compose 99% of the secondary mineral mass that precipitates in the pore space of PRBs under a broad variety of conditions. Additionally, the three minerals included in the model represent the most common minerals observed in column and field studies (Mackenzie *et al*., 1999; Phillips *et al*., 2003; Li, 2004). However, the model is sufficiently general to permit inclusion of other minerals to simulate site-specific conditions.

Natural aquifer heterogeneity was simulated with a random field approach using the turning bands method described in Elder *et al*. (2002). Hydraulic conductivity of the aquifer was assumed to be log-normally distributed and was characterized by the geometric mean hydraulic conductivity (K_g), the standard deviation of the logarithm of hydraulic conductivity ($\sigma_{\ln K}$), and correlation lengths along the principles axes (λ_l, λ_t, and λ_v). The PRB is located in the middle of the aquifer and oriented perpendicular to the primary direction of groundwater flow, as shown in Fig. 1. The PRB is 10-m deep, 25-m wide, and 1-m thick, and is assumed to have uniform hydraulic conductivity (K_{po} = 216 m day^{-1}) and porosity (0.6) at the time of installation. For the simulations described in this paper, $K_g = 3.9$ m day^{-1}, $\sigma_{\ln K} = 1.0$, $\lambda_l = 3$ m, $\lambda_t = 1$ m, and $\lambda_v = 0.5$ m. Results for a broad range of aquifer properties can be found in Li (2004).

MODFLOW (McDonald & Harbaugh, 1988) was used to simulate steady-state flow in the PRB and a portion of the surrounding aquifer. The model domain consisted of 250 columns, 120 rows, and 20 layers, which represents a flow domain of 71 × 60 × 10 m. The grid spacing was 0.5 m in the vertical and lateral directions and varied from 0.3 m to 0.1 m in the longitudinal direction, with the smaller spacing (0.1 m) being used within the PRB. Specified-head boundaries were set along the east and west ends of the aquifer to establish an average hydraulic gradient of 0.01 in the longitudinal

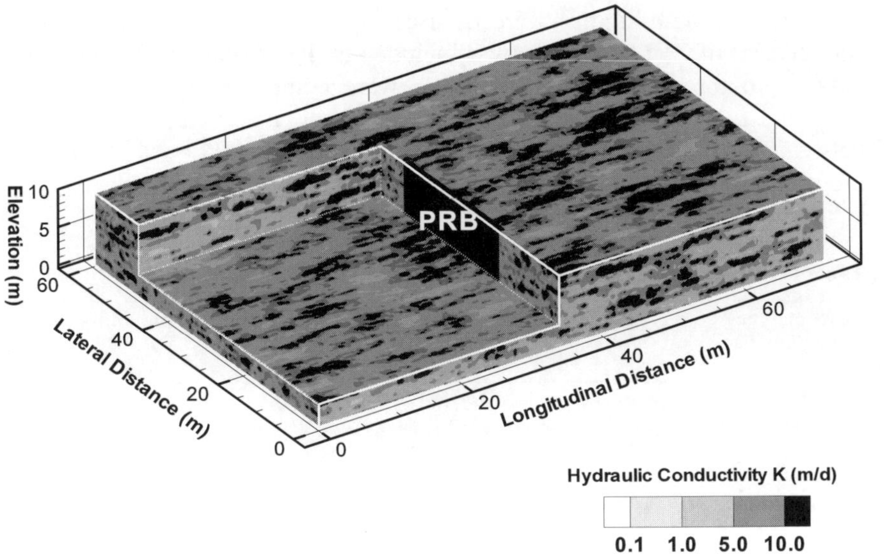

Fig. 1 Heterogeneous aquifer containing PRB that was simulated in study.

direction. No-flow boundaries were used for the lateral, top, and bottom boundaries. The flow field was updated annually to account for changes in hydraulic conductivity of the PRB caused by fouling. The frequency of updating was based on a parametric study (Li, 2004), which showed that updating more than once per year had a negligible effect on the flow field. A Kozeny-Carman approach was used to update hydraulic conductivities in the PRB (Li, 2004).

Geochemical processes within the PRB were simulated with RT3D, a three-dimensional multi-species code describing advective–dispersive–reactive transport (Clement, 1997) that uses the head solution from MODFLOW as input. The initial concentration of each species in the PRB was assumed to be zero. The upstream boundary was assigned the concentration of ions in groundwater and was assumed to be a spatially uniform and time invariant. The bottom boundary was assigned as no flux. All other boundaries were assigned a Cauchy boundary condition with no dispersive flux. Groundwater entering the PRB was assumed to be anaerobic and have the following composition: DO = 10^{-8} M, Fe^{2+} = 10^{-10} M, Ca^{2+} = 10^{-3} M, OH^- = 10^{-7} M, HCO_3^- = 10^{-3} M, and CO_3^{2-} = 10^{-7} M. Simulations for aerobic conditions can be found in Li (2004).

A geochemical algorithm developed by the authors was incorporated into RT3D for simulating the geochemical reactions in the PRB (Li, 2004). Two redox reactions related to iron corrosion and three mineral precipitation–dissolution reactions were included (Table 1). A pseudo first-order reaction rate proportional to the reactive surface area of ZVI and the DO concentration was assumed for aerobic iron corrosion (Mayer *et al.*, 2001). For anaerobic iron corrosion, the reaction rate was assumed to be proportional only to the reactive surface area of ZVI (Reardon, 1995). The kinetics of mineral precipitation–dissolution were assumed to follow transition state theory (Lasaga, 1998; Mayer *et al.*, 2001; Yabusaki *et al.*, 2001). Rate constants obtained by calibration to field studies were used as input (Li, 2004). Reduction of the reactive surface area of ZVI by fouling was included in the geochemical algorithm. Predictions made with the geochemical algorithm were compared with field data reported by Yabusaki *et al.* (2001) and Mayer *et al.* (2001). Excellent agreement was obtained between the field data and model predictions (Li, 2004). Additional details on the model can be found in Li (2004).

Table 1 Geochemical reactions in PRB.

Reaction type	Reactions	Solubility constant $\log(K_{eq})$	Molar volume (ml mole^{-1})
Aerobic Iron Corrosion	$Fe^{o} + H_2O + 0.5O_2(aq) \rightarrow Fe^{2+} + 2OH^-$	–	–
Anaerobic Iron Corrosion	$Fe^{o} + 2H_2O \rightarrow Fe^{2+} + 2OH^- + H_2(aq)$	–	–
Mineral Formation	$CaCO_3(s) \leftrightarrow Ca^{2+} + CO_3^{2-}$	–8.1	34.1
	$FeCO_3(s) \leftrightarrow Fe^{2+} + CO_3^{2-}$	–10.5	29.3
	$Fe(OH)_2(am) \leftrightarrow Fe^{2+} + 2OH^-$	–15.2	30.2

Mineral precipitates formed in the iron media were assumed to be immobile and the pore space occupied by the minerals was estimated from the molar volume of each mineral. The pore volume reduction at a given location was computed as the total volume of mineral precipitates at a location less the pore volume gained by dissolution of iron.

RESULTS AND DISCUSSION

Average and maximum porosity reductions in the PRB are shown in Fig. 2(a) as a function of distance from the entrance of the PRB for 10, 30, and 50 years after installation. Thin lines correspond to the average porosity reduction (arithmetic mean) and thick lines correspond to the maximum porosity reduction (greatest porosity reduction in all cells comprising the PRB) at a given distance from the entrance.

For all times, the average and maximum porosity reduction reaches a peak near the entrance (≈0.1 m), which is followed by a decrease and then leveling off (≈0.8 m from the entrance). Both the average and maximum porosity reductions increase over time as minerals accumulate. After 50 years, the maximum porosity reduction reaches 0.58 near the entrance and the peak average porosity reduction is approximately 74% of the pore space. Thus, at 50 years, substantial blockage of the PRB has occurred. Even at 30 years, approximately one-half of the pore space has been blocked in some portions of the PRB.

The difference between the average and maximum porosity reductions reflects the impact that flow heterogeneity has on the rate of mineral precipitation. These differences are greatest near the entrance face because mineral precipitates in this region largely consist of carbonate minerals, and the rate at which these minerals

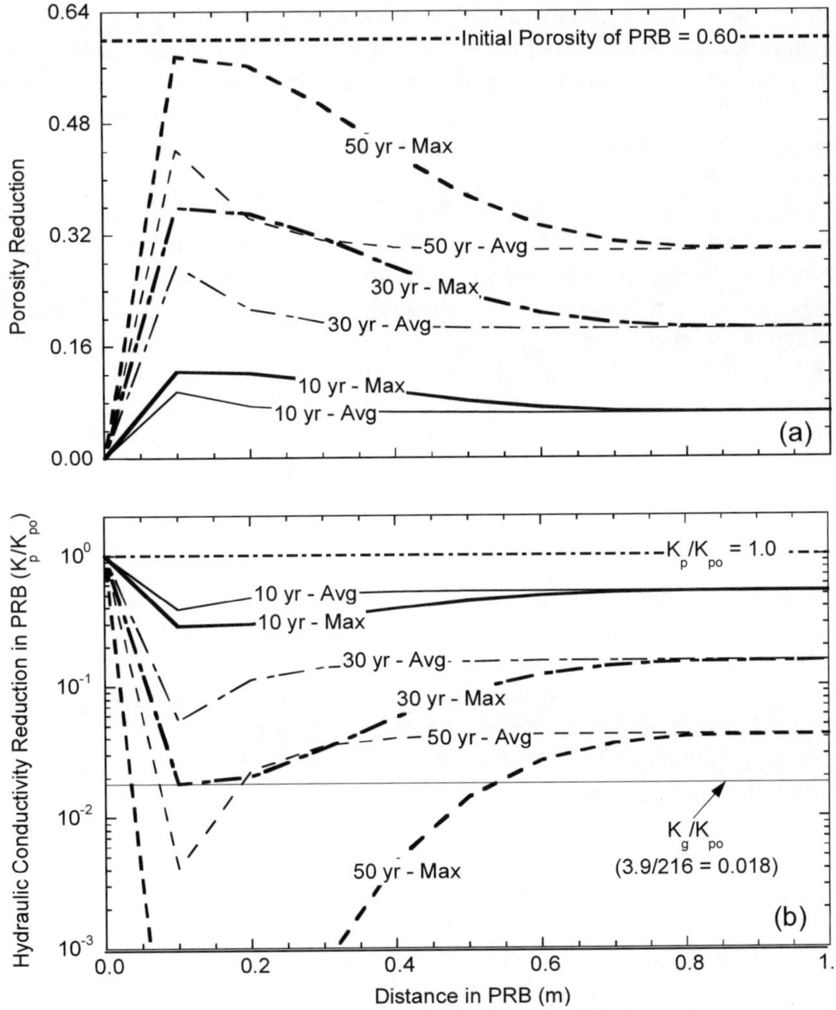

Fig. 2 Porosity reduction (a) and hydraulic conductivity reduction (b) after 10, 30, and 50 years as a function of distance from the entrance face of the PRB.

accumulate is controlled by the spatially varying rate at which mineral forming ions (primarily Ca^{2+}, CO_3^{2-}, OH^-, HCO_3^-) enter the PRB via advection. The difference between the maximum and average porosity reductions in the region between the entrance and mid-plane increases over time, which reflects the higher rate at which minerals accumulate in pores associated with preferential flow paths. In contrast, the maximum and average porosity reductions are indistinguishable near the exit of the PRB. Porosity reductions near the rear of the PRB are caused almost exclusively by ferrous hydroxides, which are formed primarily from ions generated by corrosion processes (Fe^{2+} and OH^-) within the PRB rather than ions entering the PRB via inflowing groundwater (Li, 2004).

The effect that mineral accumulation has on hydraulic conductivity within the PRB is shown in Fig. 2(b) in terms of the hydraulic conductivity ratio (K_p/K_{po}), where K_p is the hydraulic conductivity at a point within the PRB at any time and K_{po} is the initial hydraulic conductivity at that point. Thick lines correspond to the minimum K_p/K_{po} (largest porosity reductions) and thin lines correspond to the average K_p/K_{po}. When K_p/K_{po} falls below 0.018 (lower horizontal line in Fig. 2(b)), portions of the PRB are less permeable than the geometric mean hydraulic conductivity of the aquifer, a condition that may cause bypassing.

Both the minimum and average K_p/K_{po} decrease over time in response to the accumulation of minerals. Over the first 10 years, the reductions in hydraulic conductivity are modest and all portions of the PRB remain much more permeable than the aquifer. This finding is generally consistent with field observations to date, which show that properly engineered PRBs continue to conduct flow efficiently for years (Wilkin & Puls, 2003). After 30 years, some regions of the PRB are less permeable than the aquifer, but the average hydraulic conductivity of the PRB remains above the hydraulic conductivity of the aquifer. Thus, the PRB should still convey flow efficiently even after 30 years, which is greater than the service life generally expected for PRBs. However, after 50 years, large reductions in hydraulic conductivity have occurred up to the mid-plane of the PRB. These reductions in hydraulic conductivity should cause flow alterations and partial bypassing of the PRB.

Darcy velocities at the entrance of the PRB are shown in Fig. 3 at the onset of operation and after 10, 30, and 50 years. The Darcy velocities are spatially variable at all times due to the spatial variability of the aquifer hydraulic conductivity. Inspection of Fig. 3(a) and (b) indicates that regions with the highest Darcy velocities (≈0.5 m day^{-1}, shaded black) become slightly smaller during the first 10 years of operation, and those with moderate Darcy velocity (≈0.03–0.06 m day^{-1}, shaded grey) become slightly larger. This effect is more evident when the distributions at 0 and 30 years are compared (Fig. 3(a) and (c)). Regions that initially have the highest Darcy velocities no longer exist at 30 years, regions with the lowest Darcy velocities (≈0.01 m day^{-1}) have shrunk slightly, and regions with moderate Darcy velocities have increased in size. In effect, mineral precipitation in the pore space is redistributing flows from the most permeable and least permeable pathways to the pathways that are moderately permeable.

Changes that occur between 30 and 50 years are the most dramatic. At 50 years, the regions that initially had the highest Darcy velocities (and the most rapid mineral accumulation) have transitioned to regions having the lowest velocities in response to the mineral accumulation in the pore space. Similarly, many regions that initially (0 years) had moderate Darcy velocities have the highest Darcy velocities at 50 years due to the redistribution of flows in response to the changes in hydraulic conductivity.

The effect of fouling on seepage velocities and residence times in the PRB is shown in Fig. 4 using box plots. Each box plot represents the distribution (5th–95th percentiles) of seepage velocities or residence times in the entire PRB. The seepage velocities were computed using Darcy velocities obtained from MODFLOW and the corresponding porosities within the PRB, both of which account for porosity reductions by mineral precipitation. Residence times were determined using the particle tracking code Path3D (Zheng, 1991).

The seepage velocity increases because the porosity is decreasing due to fouling, whereas the Darcy velocity remains relatively constant due to the overall control by the aquifer. The gradual increase in seepage velocity over time causes a gradual reduction

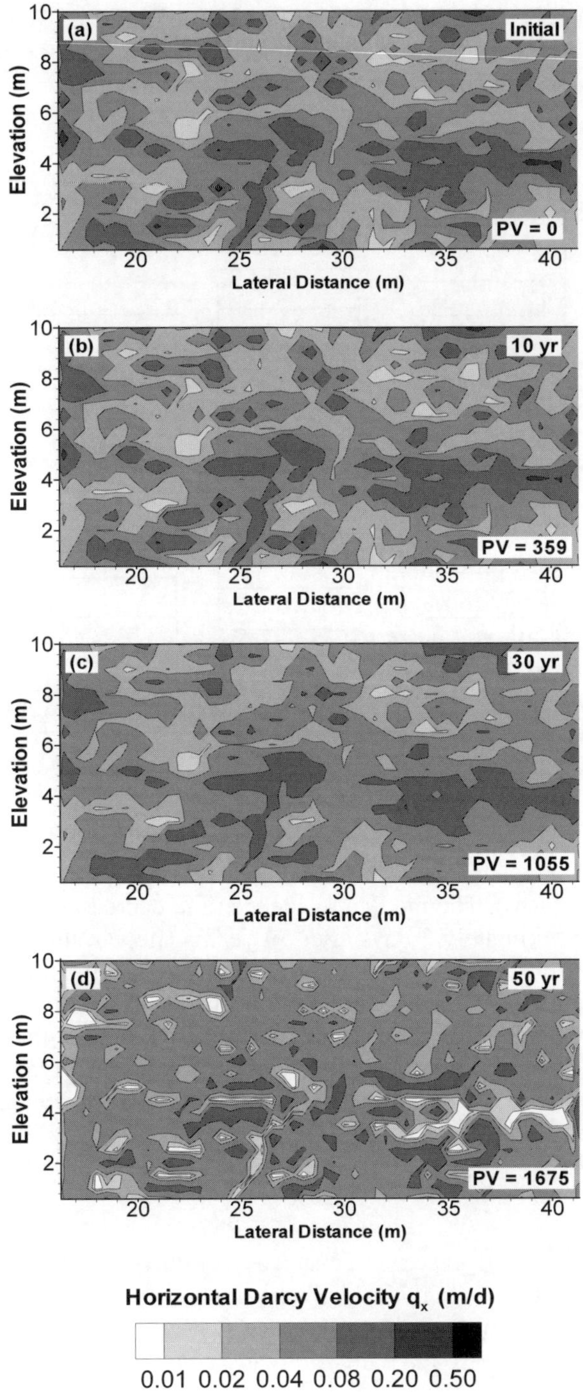

Fig. 3 Darcy velocities at the entrance face of the PRB (i.e., looking into the entrance face of the PRB): (a) initial, (b) 10 years, (c) 30 years, and (d) 50 years. PV = cumulative pore volumes of flow through the PRB.

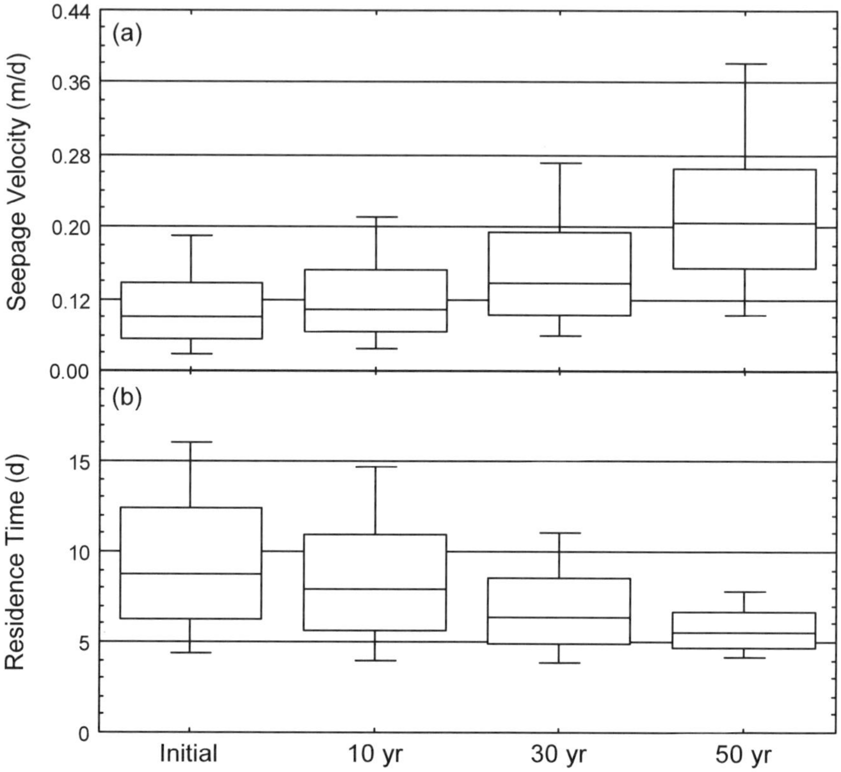

Fig. 4 Box plots of (a) seepage velocity and (b) residence time in the PRB at 0, 10, 30, and 50 years.

in residence time (Fig. 4(b)). The median residence time decreases from approximately 9 days initially to approximately 5 days after 50 years. The variation in residence time also diminishes over time (i.e. approximately 4–16 days initially and 4–8 days after 50 years). The largest reductions in residence time occur between 10 years and 50 years, i.e. outside the existing range of field experience for PRBs. The gradual reduction in residence time may have a significant effect on the long-term performance of PRBs because a shorter residence time will result in less efficient groundwater treatment.

CONCLUSIONS

A modelling study has been described that was conducted to assess the impact of mineral fouling on the long-term hydraulic behaviour of PRBs containing ZVI. Three minerals were assumed to form in the pore space ($CaCO_3$, $FeCO_3$, and $Fe(OH)_2$) and the groundwater was assumed to have the following composition: DO = 10^{-8} M, Fe^{2+} = 10^{-10} M, Ca^{2+} = 10^{-3} M, OH^- = 10^{-7} M, HCO_3^- = 10^{-3} M, and CO_3^{2-} = 10^{-7} M. The analysis shows that little change in hydraulic behaviour occurs within the first 10 years of operation, which is consistent with field experience to date for properly engineered PRBs. Under the assumed conditions, significant changes in hydraulic behaviour should be expected after approximately 30 years due to reductions in porosity and hydraulic conductivity. After 50 years, large regions of PRBs are likely to become clogged and less permeable than the aquifer, resulting in bypassing of groundwater. The magnitude of these effects is influenced by the rate at which major ions enter the PRB via advection (i.e. ionic strength and velocity of inflowing groundwater). More rapid fouling will occur at sites where conditions are prone to greater rates of mineral precipitation.

Flow redistribution should be expected during the life of a PRB. Flows through more permeable pathways should diminish because these pathways accumulate minerals at a greater rate. Nevertheless, the seepage velocity in a PRB should gradually increase because the flow rate (Darcy velocity) through a PRB generally is controlled by properties of the aquifer, which do not change appreciably over time. Over 30–50 years, increases in seepage velocity should cause reductions in residence time on the order of 50% under the assumed conditions. Larger and more rapid increases in seepage velocity may occur at sites where minerals precipitate at greater rates.

REFERENCES

Clement, T. (1997) *A Modular Computer Code for Simulating Reactive Multi-Species Transport in 3-Dimensional Groundwater Aquifers*. PNNL-SA-11720, Pacific Northwest National Laboratory, Richland, Washington, USA.

Elder, C., Benson, C. & Eykholt, G. (2002) Effects of heterogeneity on influent and effluent concentrations from horizontal permeable reactive barriers. *Water Resour. Res.* **38**(8), Art. No. 1152.

Lasaga, A. (1998) *Kinetic Theory in the Earth Sciences*. Princeton University Press, Princeton, New Jersey, USA.

Li, L. (2004) Fouling and the long-term performance of permeable reactive barriers. PhD Dissertation. University of Wisconsin-Madison, Madison, Wisconsin, USA.

Mackenzie, P., Horney, D. & Sivavec, T. (1999) Mineral precipitation and porosity losses in granular iron columns. *J. Hazardous Materials* **68**, 1–17.

Mayer, K., Blowes, D. & Frind, E. (2001) Reactive transport modeling of an *in situ* reactive barrier for the treatment of hexavalent chromium and trichloroethylene in groundwater. *Water Resour. Res.* **37**, 3091–3103.

McDonald, M. & Harbaugh, A. (1988) MODFLOW, A modular three-dimensional finite-difference ground-water flow mode. *US Geological Survey, Washington, DC, USA.*

Philips, D., Watson, D., Roh, Y. & Gu, B. (2003) Mineralogical characteristics and transformations during long-term operation of a zerovalent iron reactive barrier: mineralogical characteristics. *Environ. Sci. Technol.* **34**, 4169–4176.

Reardon, E. (1995) Anaerobic corrosion of granular iron: measurements and interpretation of hydrogen evolution rates. *Environ. Sci. Technol.* **29**, 2936–2945.

Wilkin, R. & Puls, R. (2003) Capstone report on the application, monitoring, and performance of Permeable Reactive Barriers for ground-water remediation. *EPA 600-R-03-045, US EPA, Cincinnati, Ohio, USA.*

Yabusaki, S., Cantrell, K., Sass, B. & Steefel, C. (2001) Multicomponent reactive transport in an in situ zero-valent iron cell. *Environ. Sci. Techno.* **35**, 1493–1503.

Zheng, C. (1991) *Path3D: A Ground-water Path and Travel-time Simulator.* S. S. Papadopulos and Assoc., Bethesda, Maryland, USA.

Permeable Reactive Barriers (Proceedings of the International Symposium held at Belfast, Northern Ireland, March 2004). IAHS Publ. 298, 2005.

The first commercial permeable reactive barrier composed of granular iron: hydraulic and chemical performance at 10 years of operation

SCOTT D. WARNER[1], BETTINA L. LONGINO[1], MIAO ZHANG[1], PETER BENNETT[1], FRANK S. SZERDY[1] & LISA A. HAMILTON[2]

1 *Geomatrix Consultants, Inc., Oakland, California, USA*
swarner@geomatrix.com

2 *General Electric Company, King of Prussia, Pennsylvania, USA*

Abstract The year 2004 marks the tenth year of operation for the first commercial permeable reactive barrier (PRB) composed of granular iron metal. Recent performance assessment techniques, including the use of tracer dilution testing, analysis of groundwater samples for dissolved gases (including hydrogen and methane) and inorganic constituents have shown that the system continues to function successfully after nearly 10 years since its installation. The PRB, which is intended to reduce the concentration of aliphatic volatile organic compounds in groundwater to state water quality standards, consists of a 1.3-m thick (flow-through distance) by 10-m long by 6-m deep zone of granular iron; the 1.3-m thick granular iron zone is sandwiched on its up- and down-hydraulic gradient (laterally) sides by 0.7-m thick mixed iron/gravel zones. Cement-bentonite and soil-cement-bentonite slurry walls extend up to 100 m upgradient from the PRB to contain and route groundwater through the treatment area. Assessment of the PRB has utilized both chemical and hydraulic methods intended to evaluate not only the current conditions of the PRB, but to provide information pertinent to projecting future long-term performance. The evaluation of the analytical results indicates that at 10 years, groundwater within the iron core of the PRB is: (a) limited in biological activity; (b) strongly reducing (–500 mV as equivalent Eh), high in pH (11 pH units) and low in calcium and magnesium compared to ambient groundwater outside the PRB; and (c) high in dissolved hydrogen with concentrations as high as 600 000 nano-molar (nM), which is close to the aqueous solubility of hydrogen under standard conditions. A geochemical transition zone (including moderate hydrogen concentration, moderately reducing conditions, elevated pH, and higher methane concentration relative to influent groundwater) continues to occur within the upgradient iron/gravel zone. The results of tracer dilution testing in wells installed in the PRB suggest that groundwater velocity within the PRB system ranged from approximately 0.07 to 0.2 m per day during the testing. These velocity values are generally consistent with expectations 10 years ago where a design velocity of approximately 0.16 m per day was anticipated. Based on the current performance results, the PRB continues to meet the general intent of the design developed 10 years ago.

Key words commercial installation; hydraulic and chemical performance; permeable reactive barrier

INTRODUCTION

The permeable reactive barrier (PRB) was first pilot-tested at the Canadian Air Force Base Borden site as an *in situ* groundwater treatment method in 1991 by researchers from the University of Waterloo (Gillham & O'Hannesin, 1994; O'Hannesin & Gillham, 1998). The first commercial application of the PRB as a final groundwater remedy came three years later with the installation of the PRB at a former manufacturing site in Santa Clara County, California, USA (Yamane *et al.*, 1995; Szerdy, *et al.*, 1996; Warner & Sorel, 2003). Since that time, PRBs have been reportedly installed at more than 80 sites on four continents as full-scale groundwater remedies to treat a variety of organic and inorganic chemical constituents.

The subject site remains the longest running commercial application of the PRB remediation method. The PRB alternative was first conceptualized for the Sunnyvale site in 1991 as a potential alternative to replace an interim groundwater extraction and treatment system for treating chlorinated aliphatic compounds in groundwater beneath the former semiconductor manufacturing facility. Following nearly three years of design work (including laboratory and field pilot testing) and regulatory discussions, construction of this PRB began in November 1994. The former groundwater extraction and treatment system at the site was decommissioned following completion of the PRB leaving the site with no aboveground evidence of a functioning remediation system, with the exception of a number of monitoring well plates remaining at ground level.

At nearly 10 years of operation, interest in site-specific assessment of the current and future performance of this PRB is high. System monitoring to demonstrate regulatory compliance is performed quarterly to semi-annually; a comprehensive performance assessment occurs at approximately five-year intervals. During these performance assessments, evaluation of the hydraulic and geochemical conditions is used to assess current performance, as well as future viability of the system. The overall costs of the remedy compared with the long-term operation and maintenance associated with the former groundwater extraction and treatment system are also reviewed. A summary of several components of the 10-year performance assessment is provided in this paper.

Site layout and PRB installation

The site plan and monitoring well network for this PRB project is depicted by Fig. 1. Hydrogeologic investigations have identified two water-yielding zones at the site and vicinity which are referred to as the shallow groundwater zone and the deeper groundwater zone. The shallow groundwater zone is encountered generally at a depth of 3–5 m below ground surface (b.g.s.) and generally consists of a semi-confined system composed of interbedded silt, sand, and clay. The zone generally is less than 1 m to about 1.3 m thick across the site, but can be as thick as 3.3 m thick locally. Transmissivity of the zone was calculated to be approximately 27 m^2 day^{-1} during site characterization activities in the early 1990s. The shallow groundwater zone is separated from the deeper regional groundwater system by approximately 20 m of fine-grained sediment (e.g. clay and silt). Water level measurements indicate that the direction of the vertical hydraulic gradient between the shallow and deeper zone consistently is upward.

Groundwater levels in wells and piezometers within the PRB and the shallow groundwater system range from approximately 1.3 to 3 m below ground surface; water levels fluctuate up to about 0.7 m seasonally in response to winter precipitation. The regional direction of flow indicated by these water levels is generally northward

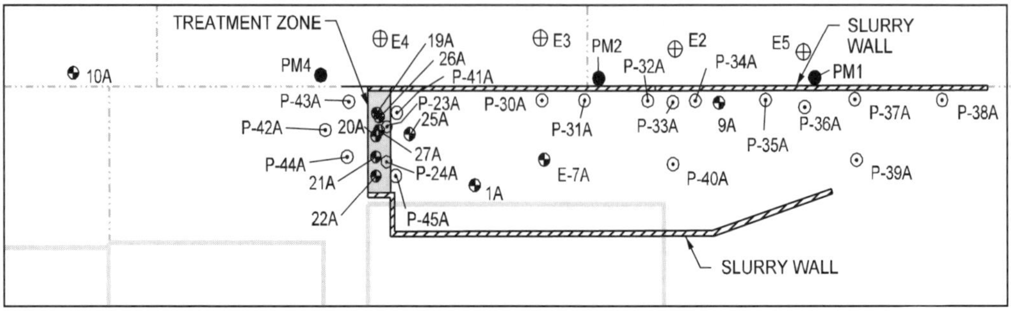

Fig. 1 Layout of the PRB system with monitoring well locations indicated (approximate scale: 2.5 cm equals 20 m. Flow direction is from right to left.

toward San Francisco Bay. In the site vicinity, the groundwater flow has been observed to vary from northwest to northeast. The regional horizontal flow gradient is approximately of 0.004 m min^{-1}.

The PRB system consists of two main components: (1) a reactive cell approximately 10 m long by 6 m deep by 2.3 m wide and containing a 1.3 m wide flow-through zone of 100% granular iron sandwiched between 0.7-m wide upgradient and downgradient sections of pea gravel; and, (2) low permeability soil-cement-bentonite subsurface cutoff walls extending nearly 100 m upgradient from the treatment cell (Fig. 1). A sheet pile wing wall also extends upgradient approximately 6 m from one end of the treatment cell. The low permeability walls were installed to contain the site and reduce the potential impact of variable ambient groundwater flow directions. The treatment cell is keyed into the lateral low permeability barriers (Fig. 2).

The PRB zone was constructed by excavating out the PRB cell between sheet piles that had been installed to delineate the treatment cell. Monitoring wells consisting of 5-cm diameter PVC were placed in the granular iron within approximately 0.3 m of its downgradient and upgradient limits and within the centre of the granular iron section. Following the backfilling activity, the sheet piles were removed; the action of

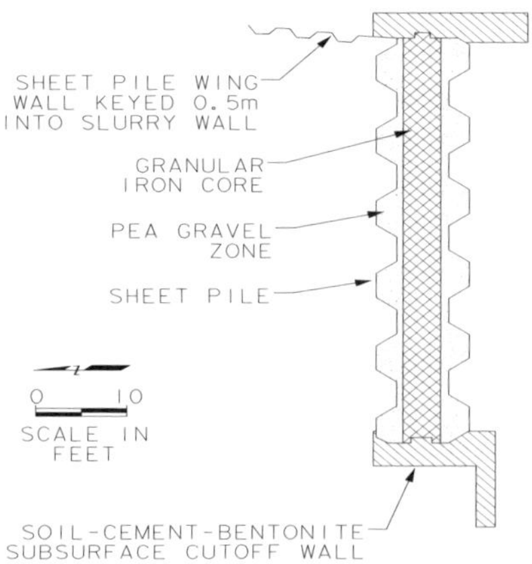

Fig. 2 Depiction of the keying of the treatment cell with low permeability cut-off walls (10 feet ~ 3 m).

removing the sheet piles likely caused some mixing of granular iron and pea gravel in the up- and downgradient compartments.

Remediation intent and process

The PRB is intended to reduce the concentration of target chlorinated aliphatic compounds in groundwater flowing across the site's northern (downgradient) boundary to below California water quality standards. These target compounds consist chiefly of trichloroethene (TCE) and its daughter products, including cis-1,2-dichloroethene (*cis*-1,2-DCE), and chloroethene; the California water quality standard is 5.0 μg l^{-1} for TCE, 6.0 μg l^{-1} for *cis*-1,2-DCE, and 0.5 μg l^{-1} for chloroethene. The fluorinated ethane compound 1,1,2-trichloro-1,2,2-trifluoroethane (Freon 113) also is considered a target compound. The treatment process (patented by researchers from the University of Waterloo, Ontario, Canada) involves the use of granular iron to promote chemical destruction of the target compounds. The chemistry involved in the treatment process has been well studied and published by others (for example, see other papers in this volume) and will not be presented in detail here.

Several secondary hydrochemical effects are related to the treatment process occurring within the PRB. Typical effects include: increased pH, lowered reduction-oxidation (redox) potential, consumption of dissolved oxygen, shift of carbonate equilibria (in response to the pH change), formation of solids, and evolution of hydrogen and low-molecular weight hydrocarbon gases such as methane.

Hydraulically, the system is not intended to affect the ambient flow system aside from assuring that affected groundwater upgradient of the PRB is routed through the treatment cell and remains within the cell at sufficient resident time to allow complete treatment. However, the intrusive nature of the PRB as well as its design as an unconfined permeable trench within a semi-confined aquifer system does influence the flow system.

The following sections provide performance related data for the chemical and hydraulic conditions of the PRB as the system approaches its tenth year of operation.

Hydrogeochemical evaluation

Compliance monitoring of this PRB primarily involves determining the concentration of target chemicals within groundwater samples collected from monitoring wells located within the PRB, although other parameters and analytes are monitored and reported to the lead regulatory agency. Performance of the PRB, however, is better assessed by combining information from the compliance monitoring programme with an evaluation of additional parameters and analytes that provide functional information on the PRB process.

Compliance monitoring summary

Since installation of the PRB, the compliance monitoring programme has involved routine sampling and analysis of groundwater from monitoring wells for 13 chemical analytes, and monitoring water levels from project wells, as ordered by the lead regulatory agency. The frequency of data collection has evolved from monthly during the first few months following installation, to quarterly for several years, to a combination of quarterly and semi-annual. Early groundwater samples from within the PRB well network were collected using low-flow or micro-purge techniques; current methods involve the use of passive diffusion bags (PDB) to provide samples for assessing the concentration of the target chemicals in groundwater. The PDBs are placed and retrieved from the PRB at approximately six-month intervals; a pilot test of the PDB use for this site has bcen described by others (Sorel *et al.*, 2003) .

Figure 3 provides a representative summary of the concentration of target analytes in groundwater samples collected from monitoring wells located upgradient, within, and downgradient of the PRB. Generally, target analyte concentrations are consistently lower within the PRB than in samples from either upgradient or downgradient wells. Groundwater samples from wells installed within the transitional upgradient pea gravel zone, though not part of the compliance monitoring programme, show a transitional concentration of analytes indicating some affect from either the small amount of iron mixed within the pea gravel or an upgradient influence from the granular iron core of the PRB. Low concentrations of some target analytes, including vinyl chloride, however, are found in samples from some wells (e.g. well 19A in Fig. 1) located at the downgradient edge and one end of the PRB cell (within the last 0.2 m of the granular iron zone). It is believed that a chemical source exists offsite just downgradient and adjacent to the PRB cell, and that this may be a source of the low concentrations observed in samples from well 19A (well 19A is located at the north and intended downgradient edge of the PRB where, if chemicals migrated into this well from the north, the affected water may have had insufficient residence time for complete treatment). Target analytes have never been detected in samples from wells in the centre of the PRB (e.g. wells 26A and 27A on Fig. 1) where the residence time is designed for complete treatment to occur. A short period of groundwater flow reversal (from north to south, rather than the ambient south to north—see Hydraulics section) caused by construction and dewatering activities at an adjacent site likely affected the observed chemical conditions.

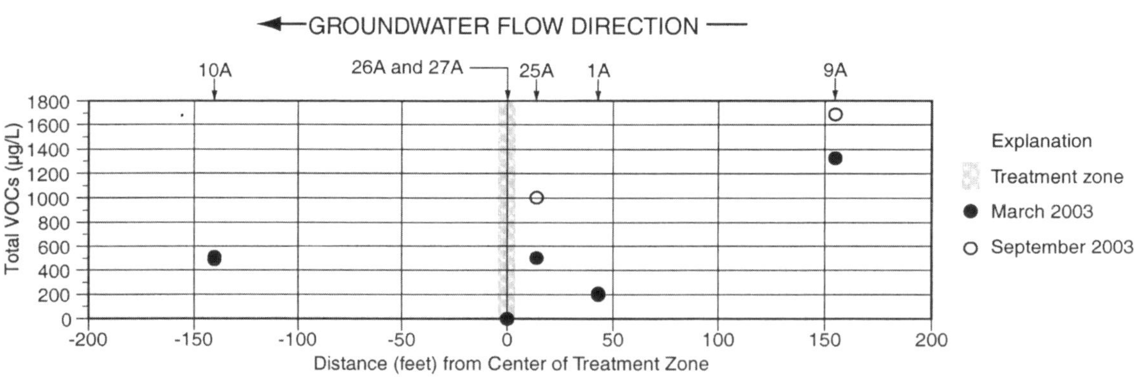

Fig. 3 Representative concentration of total VOCs in monitoring wells within, upgradient, and downgradient of the PRB treatment cell.

Performance monitoring summary

Parameters used to assess PRB peformance include pH, redox potential, dissolved oxygen (each measured routinely), and the concentration of inorganic analytes, dissolved hydrogen, and hydrocarbon gases, and biological analytes. The following bullets summarize the monitoring results for selected parameters at the 10-year mark.

pH The pH of ambient groundwater is near neutral (7–7.5) while pH values for groundwater samples within the pea gravel are near 9.5 and within the PRB core range from about 10 to 12 pH units. Downgradient values return to near neutral. A slight drift to higher pH values over time appears to be occurring within the PRB core (Fig. 4).

Redox potential Field values are measured as oxidation-reduction potential using a voltmeter with a platinum electrode. Values, converted to Eh, indicate strong reducing conditions through the 10-year history of the PRB with consistent values of about –300 to –500 mV (Fig. 5).

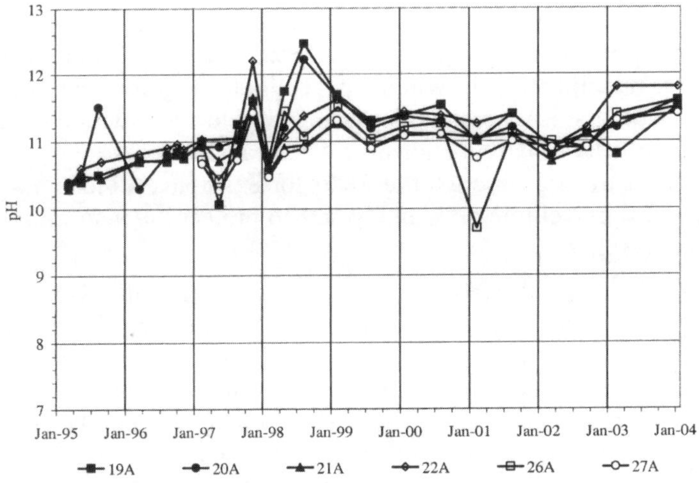

Fig. 4 pH measurements with time for treatment zone wells.

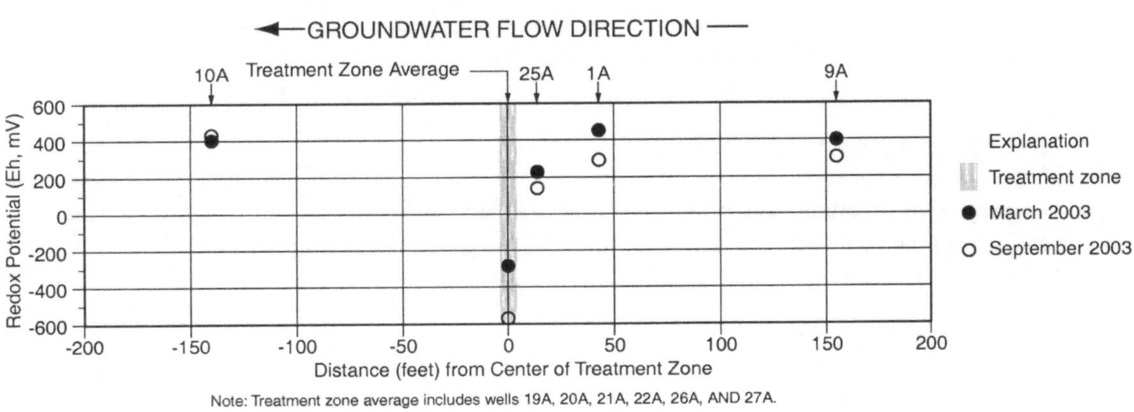

Fig. 5 Relation of redox potential along flow in the PRB system.

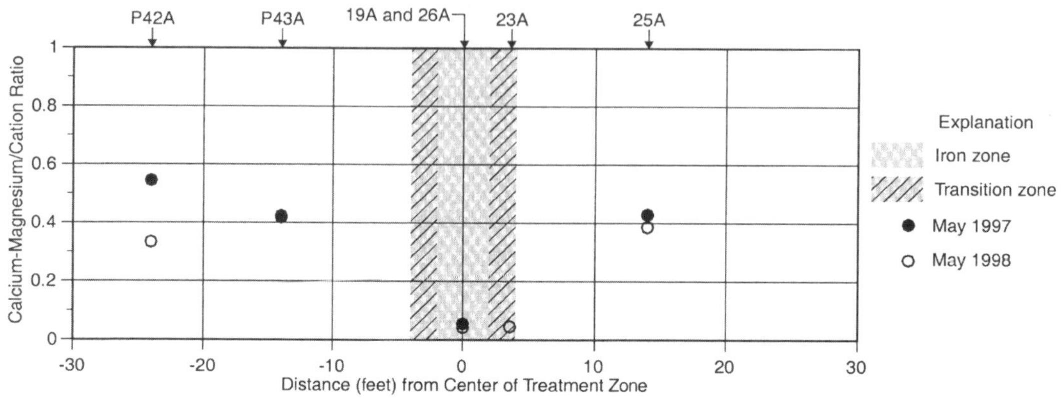

Fig. 6 Relation of the Ca-Mg/cation ratio along flow in the PRB system.

Inorganic analytes Concentrations of major inorganic constituents, including sodium, calcium, magnesium, potassium, bicarbonate, chloride and sulphate, indicate that solids may be forming within the pore space of the PRB treatment cell. The most likely area for precipitation is within the upgradient gravel/iron section where the initial pH shift occurs; less precipitation is believed to occur within the central granular iron core of the PRB. The reduction in calcium as a ratio of total cations in samples collected along a transect through the PRB, for example, is illustrated by Fig. 6. The upgradient mixed gravel/iron zone likely acts to protect the main iron core from rapid accumulation of solids.

Hydrogen and hydrocarbon gases Concentrations of dissolved hydrogen and other hydrocarbon gases (such as methane) have been measured on multiple occasions during the 10-year life of the PRB. Methane and other hydrocarbon gases were measured first in 1997 and then again in 2002. The data suggest that near solubility levels (solubility represented as approximately 800 000 nanomolar (nM) or 1.8 mg l^{-1}) of dissolved hydrogen were measured within the granular iron core more than six years after the PRB was installed; these values were only slightly reduced in 2002, nearly nine years after installation. The occurrence of methane continues with higher concentrations in the upgradient pea gravel section of the PRB. Increased concentrations of hydrogen with respect to expected natural conditions are measured some 3 m upgradient of the PRB in ambient well 25A.

HYDRAULIC EVALUATION

The following sections summarize and interpret hydraulic information collected for the PRB compliance and performance monitoring programmes. Data is presented in terms of water level information, and tracer testing results.

Water level measurements

Figure 7 provides a history of water level measurement for wells within the PRB system over the past 10 years. Several consistent conditions are apparent: (a) water

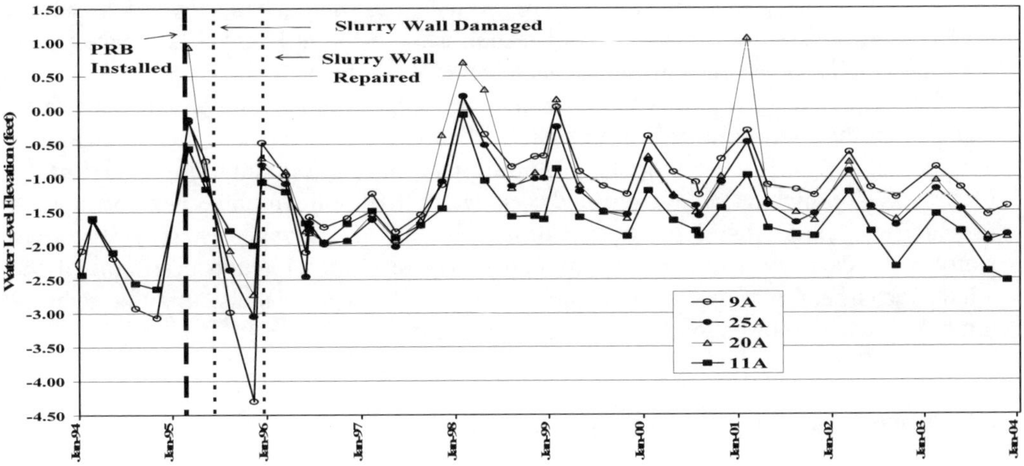

Fig. 7 Ten-year histograph of water levels in monitoring wells located within the PRB cell.

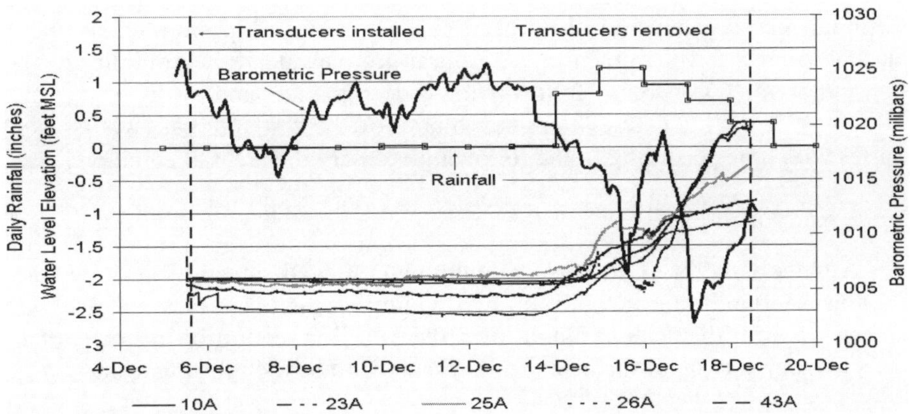

Fig. 8 Pressure transducer readings for the PRB cell during an initial recharge event.

levels typically increase in the winter and decline following the winter rainy season; (b) water levels appear to be higher within the PRB trench than in surrounding wells during the winter recharge season but also decline following the rainy season; and (c) regional water levels increased greatly following the winter of 1997/1998 and have gradually declined since then.

Also of note are two particular hydraulic events for this PRB. The first is indicated by the notable decline in water levels in the 1995–1996 time period. This decline, which led to a reversal of the horizontal hydraulic gradient (backward flow through the PRB), resulted from a combination of dewatering and breach of the protective upgradient and eastern low permeability barrier wall during construction activities on the site adjacent to the eastern property line. Grout injection repaired the breach and water level conditions returned to expected conditions.

Water levels in wells within the PRB are consistently greater than in wells outside the PRB during periods of recharge; however, these water levels decline rapidly during the dry season to levels that are consistent with the intended northward hydraulic gradient. An evaluation of water level response to recharge event was conducted in 2002 by placing pressure transducers in selected wells located within and external to the PRB. Figure 8 shows the results of this monitoring event, during which water level response to the first major rain storm and recharge event of the season was measured and recorded. After the development of an initial transient hydraulic depression in the PRB trench (which is an unconfined system), a hydraulic mound developed with respect to the external system (semi-confined shallow groundwater system). Historical water level data confirm that this mound dissipates following the cessation of the winter recharge. Groundwater flow through the PRB during recharge may be reduced until the gradient returns to non-recharge conditions.

Assessment of ambient flow velocity through the PRB

Tracer dilution tests were conducted in PRB monitoring wells in October 2003. Tracer dilution tests involve the addition of a "tracer" to the screened interval of a well, followed by monitoring of the concentration of the tracer over time in the same well. Periodic measurements of the tracer concentration are performed as the tracer is flushed from the well screen under natural groundwater flow conditions. The rate of concentration (C) decrease follows first-order kinetics, and is directly related to the dilution rate (Q/V); where V is the volume of the test interval (well-screen) and C varies with time according to the following ordinary differential equation:

$$\mathrm{d}C/\mathrm{d}t = -\left(Q/V\right) \cdot C(t) \tag{1}$$

After solving for Q, the flow rate through the well-screen (Q) can be converted to the linear groundwater velocity (v) by dividing by the cross-sectional area of the well screen (A; well diameter × length of screen), a flow distortion factor, alpha, and the effective porosity, n, of the iron media (assumed to be 0.40 for this work):

$$v = Q/(nA\alpha) \tag{2}$$

The average groundwater flow rate (i.e. the flow rate for the duration of the test) through the screened intervals of the tested monitoring wells ranged from approximately 4 to 15 ml min^{-1}, with the highest flow rate measured at the west end of the PRB. Using equation (2), the calculated ambient groundwater velocity through the treatment zone ranged from approximately 0.05 to 0.2 m day^{-1}. These results were used to calculate a range of hydraulic residence time for the PRB of approximately 7–22 days. Past laboratory and field studies conducted for the treatment wall estimated that the minimum residence time required for treatment of vinyl chloride to regulatory cleanup standards is approximately 2 days.

System cost to-date

When the PRB remedy was first conceptualized, affected groundwater at the site was subject to pump-and-treat remediation. Three to four extraction wells were pumping an average of 3–5 l min^{-1} to a treatment plant consisting of an air stripper for primary treatment and granular activated carbon for secondary treatment. The cost components of that remedy included operation and maintenance of the system, groundwater monitoring, groundwater compliance reports, effluent discharge reports, and various legal and regulatory activities. Also, the party responsible for the environmental conditions was required to maintain an expensive lease with the landowner, even though the primary party's business had not been active for several years. With the implementation of the PRB, after the capitalization of the project, much of the prior operating cost was eliminated, including effluent discharge reports and the property lease (with no above ground treatment system, the owner allowed a new tenant into the property within five months following completion of the PRB construction). Figure 9 illustrates the relative cost savings by comparing relative cost conditions from actual operation relative to the projected costs that would have occurred had the former pump and treat system continued to operate during the past 10 years; a comparison with projected relative costs for the PRB remedy when first conceptualized is also provided.

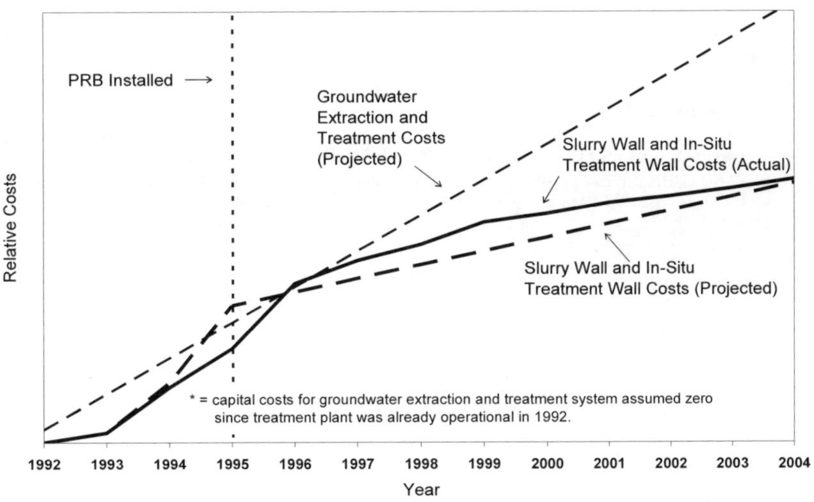

Fig. 9 Relative costs for the PRB remedy compared to predicted costs and the former remedy.

CONCLUSION

Approaching 10 years of operation, this first commercial application of a granular iron PRB has achieved its primary intent: to provide a regulatory acceptable groundwater remedy for affected groundwater at the site for a cost lower than the previous conventional remedy while allowing full economic utilization of the property to occur. Technical data indicate that both geochemical and hydraulic conditions still require regular monitoring and assessment, and it seems likely that the PRB will remain effective beyond the 10-year time frame. Key observations from the nearly 10 year operation include:

(a) The mixed upgradient pea gravel/iron zone is a hydrochemical transition zone that protects the central granular iron treatment cell.
(b) The chemical treatment process continues to function as designed many years after system installation.
(c) The influence of the PRB goes beyond the constructed boundaries of the PRB based on the dissolved hydrogen concentration values in ambient groundwater beyond the PRB.
(d) Hydraulic conditions within the PRB fluctuate with seasonal recharge conditions.
(e) The apparent groundwater flow velocity within the PRB appears consistent with the intent of the design.
(f) The PRB remedy has been economically successful for nearly 10 years at the site.
(g) It would have been useful to have a discussion on longer term performance, based on 10 years results (as alluded to in the Abstract ".. but to provide information pertinent to projecting future long-term performance").

REFERENCES

Gillham, R. W. & O'Hannesin, S. F. (1994) Enhanced degradation of aliphatics by zero-valent iron. *Ground Water* **32**(6), 958–967.

O'Hannesin, S. F. & Gillham, R. W. (1998) Long-term performance of an *in situ* "iron wall" for remediation of VOCs. *Ground Water* **36**(1), 164–170.

Sorel, D. S., Warner, S. D., Longino, B. L., Honniball, J. H. & Hamilton, L. A. (2003) Chlorinated solvent and DNAPL remediation. *American Chemical Society, Symposium Series* **837**, 278–285.

Szerdy, F. S., Gallinatti, J. D., Warner, S. D., Yamane, C. L., Hankins, D. A. & Vogan, J. L. (1996) *ASCE National Convention*, 245–256. American Society for Civil Engineers, Washington, DC, USA.

Warner, S. D. & Sorel, D. S. (2003) Chlorinated solvent and DNAPL remediation. *American Chemical Society, Symposium Series* **837**, 36–50.

Yamane, C. L., Warner, S. D., Gallinatti, J. D., Szerdy, F. S., Delfino, T. A., Hankins, D. A. & Vogan, J. L. (1995) *209th National Meeting*, American Chemical Society, Anaheim, California, **35**(1), 792–795.

2 New Developments in PRB Technology

Comparison of different multibarrier concepts designed for treatment of groundwater containing mixed pollutants

L. BASTIAENS, J. DRIES, J. VOS, Q. SIMONS, M. DE SMET & L. DIELS

Vito, Flemish Institute for Technological Research, B-2400 Mol, Belgium

vito@vito.be

Abstract A study was set up to develop and evaluate the possibilities of multibarriers for the remediation of groundwater contaminated with mixed pollutants. A multibarrier is a multifunctional permeable reactive barrier in which biological and physicochemical pollutant removal processes can be combined. Different combinations of biodegradation, reductive dehalogenation using zero valent iron and sorption were considered to treat mixed pollutants consisting of heavy metals (As, Zn), aromatics (benzene, toluene, *m*-xylene) (BTX) and volatile chlorinated compounds (VOCls, e.g. PCE, TCE). Biodegradation of BTX in batch-systems was found to be negatively influenced by the presence of zero valent iron and other pollutants. Starting from the same aquifer mixture, addition of different electron-acceptors were selected for a different BTX-degrading microbial population as determined by PCR-DGGE. Based on column experiments which simulated different multibarrier concepts, sequential as well as mixed (Fe^0/bio) multibarriers proved to be suitable for the treatment of the tested pollutant mixture. Good results were obtained with O_2, NO_3^-, SO_4^{2-}, Fe^{3+} as terminal electron acceptors.

Key words biodegradation; mixed polllution; permeable barrier; zero valent iron

INTRODUCTION

In order to remediate contaminated groundwater, many techniques have been developed and applied. However, groundwater is often polluted with complex mixtures of different chemicals and most remediation techniques only deal with one or a few pollutant types. Treatment of groundwater polluted with mixtures of hazardous compounds has received less attention and remains a problem. A combination of different existing technologies may be a solution, but one should consider: (a) the impact of one remediation technique on another, and (b) the influence of co-contaminants on the removal efficiency of the processes.

This study focuses on multibarriers, i.e. permeable reactive barriers in which different pollutant removal processes (biological and physicochemical processes) are combined to treat *in situ* groundwater containing mixed pollutants (Fig. 1). Different multibarrier concepts to treat groundwater were designed, evaluated and compared. One of the objectives of the study was to answer the question whether the removal processes should be applied one after the other (sequential multibarrier), or whether a combination of different processes in one zone (mixed multibarrier) is also possible (Dries *et al.*, 2002, 2003). The latter may require more optimization but the installation

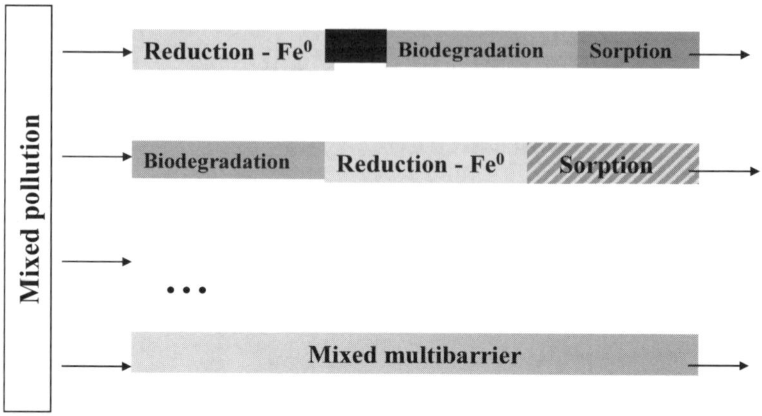

Fig. 1 Examples of multibarrier concepts.

Table 1 Summary of BTX degradation activity during the enrichment experiment: B/T/X, significant benzene, toluene or *m*-xylene degradation: 0, no degradation; N/D, not determined.

TEAP[1]	Benzene only		BTX mixture[2]		Mixed pollution	
	$-Fe^{0}$♂	$+Fe^{0}$♀	$-Fe^{0}$♂	$+Fe^{0}$♀	$-Fe^{0}$♂	$+Fe^{0}$♀
Denitrification	B	N/D	BTX	N/D	N/D	N/D
Iron reduction	B	0	BTX	TX#	TX	TX#
Sulfate reduction	B	0	TX	TX#	TX	TX#
Methanogenesis	N/D	N/D	TX	TX#	TX	TX#

[1]Terminal electron acceptor process; ♂ 45 g mixed aquifer + 30 g sand; ♀ + 150 ml medium; + 15 g Fe^0 FeA4.

[2]Benzene, toluene and *m*-xylene at 2 mg l^{-1} each; *BTX at 2 mg l^{-1}, PCE (2 mg l^{-1}), TCE (5 mg l^{-1}), Zn (5 mg l^{-1}) and As (0.2 mg l^{-1}).

#Degration halted due to pH increase.

is expected to be less complex and less expensive. Another important question concerns the choice of the terminal electron acceptors (TEA) for the biological process.

METHODS AND RESULTS

Batch enrichment experiments

A mixture of 13 different BTX, VOCls and/or heavy metal contaminated aquifer samples was used as starting material for the enrichment of benzene (B), toluene (T) and *m*-xylene (X) degrading bacteria under denitrifying, sulfate reducing, iron reducing and methanogenic conditions. The enrichments were performed in closed 250-ml vials containing 45 g mixed aquifer, 30 g filter sand and 150 ml medium. The effect of zero valent iron (Fe^0) and other pollutants (PCE, TCE, Zn, As) on the enrichment of anaerobic benzene and BTX-degrading bacteria was evaluated.

Table 1 summarizes the observed BTX degradation potential over 300 days of incubation under the different enrichment conditions. Toluene and *m*-xylene were rapidly degraded in all enrichment sets without Fe^0, and the degradation rates increased after repeated injection of new substrate. In the presence of Fe^0 the degradation halted after less than 100 days. As Fe^0 and hydrogen was not found directly toxic to relevant axenic BTX-degrading strains (Dries. 2004), the observed pH increases up to pH 8.5 may be responsible for the observed inhibition.

Benzene persisted under methanogenic conditions and in the presence of Fe^0. With nitrate, iron and sulfate as TEA degradation of benzene was observed. However, the degradation was found inhibited in the presence of PCE, TCE, Zn and As, and also in the sulfate reducing enrichments when toluene and *m*-xylene were present. The BTX-degradation capacity remained active after several (10%) transfers.

The evolution of the microbial population in the denitrifying, sulfate reducing and iron reducing enrichment cultures and the transfer cultures was followed by PCR-DGGE using general eubacterial primers amplifying a 450 fragment of the 16S rRNA gene (63F & GC-518R). Especially after a few transfers the complexity of the obtained DGGE DNA patterns decreased and a few dominant bands appeared in the sulfate and iron reducing enrichments. On the other hand, the observed population remained rather constant and complex in the denitrifying transfers. The obtained PCR-DGGE DNA patterns were different depending on the added TEA, indicating that the added TEA selected for a different population (results not shown). The dominant bands in the different DGGE-patterns have been cloned and are being sequenced.

Simulation of different multibarrier concepts in laboratory scale column systems

Potentially interesting sequential and mixed multibarrier concepts depending on the TEA are given in Fig. 2. As oxygen and nitrate are known to be transformed rapidly by zero valent iron and to have a negative influence on the performance of zero valent iron, only sequential multibarrierS (Fe^0 + BIO) were found potentially interesting with these TEAs.

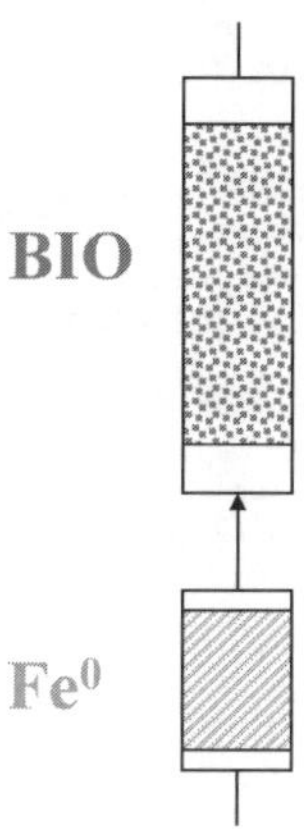

TEA	SEQUENTIAL	MIXED
O_2	✓	
NO_3^-	✓	
Fe^{3+}	✓	✓
SO_4^{2-}	✓	✓
CO_2	✓	✓

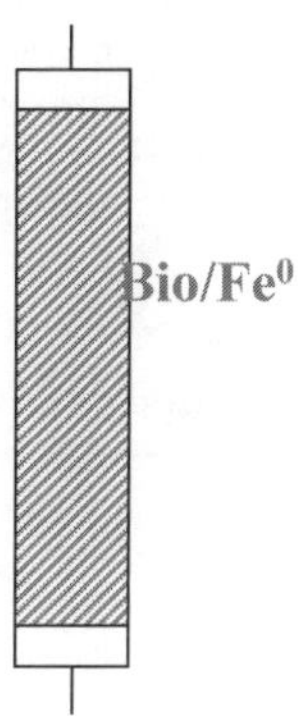

Fig. 2 Potentially interesting multibarrier concepts.

Column tests were set-up to evaluate some of these sequential and mixed multibarrierS (Table 2) and were monitored during more than 2 years (Dries *et al.*, 2004). A model pollutant mixture was defined consisting of (a) the heavy metals zinc (5 mg l^{-1}) and arsenate (0.2 mg l^{-1}), (b) the VOCls PCE (2 mg l^{-1}) and TCE (5 mg l^{-1}), and (c) the aromatic hydrocarbons benzene, toluene and m-xylene (BTX, 2 mg l^{-1} each). The artificial groundwater used in the tests contained 0.5 mM $NaHCO_3$, 0.5 mM $KHCO_3$, 0.5 mM $CaCl_2.6H_2O$ and 0.5 mM $MgCl_2.6H_2O$. The investigated pollutant removal processes were reductive dehalogenation of the VOCls with zero valent iron, sorption/reduction of the metals and biodegradation of BTX and also some VOCls. In the biobarrier oxygen (system A), nitrate (system B), sulfate (system D) as well as iron (system E) were added as the terminal electron acceptor (TEA). In all systems, the sorption part was considered mainly as a polishing step. In one column (system C) no TEA was added leaving the Fe(III) generated during corrosion of the zero valent iron or CO_2 as the only TEAs.

Table 3 summarizes the removal efficiencies of the organic compounds obtained during a certain phase in the experiment. In the sequential multibarrier simulated in systems A and B for more than 80% of the chlorinated compounds PCE and TCE were removed in the Fe^0 column, while the major part of the BTX compounds were degraded in the biocolumn. In the micro-aerobic system the removal was improved by a higher dissolved oxygen concentration.

More than 95% of the PCE and TCE were removed in the mixed zone of system C. Of the BTX compounds, 20–50% were biodegraded as determined by comparing the results of the test column train with a formaldehyde-treated set-up.

Table 2 Characteristics of different multibarrier laboratory scale column systems.

Column trains	Column characteristics (TEA)	ID	Fe (0) (w%)	FS (w%)	Aquifer (w%)	HRT	Remarks
System A: Zero valent iron wall followed by a micro-aerobic biobarrier							Aquifer = aquifer from near landfill TEA: Oxygen supplied as H_2O_2
A1	1 – Fe(0)column	A1–1	100	–	–	13 h	
	2 – Bio column (O_2)	A1–2	–	80	20	13 h	
System B: Zero valent iron wall followed by a denitrifying biobarrier							Aquifer = MB–aquifer enriched under denitrifying conditions TEA: 1 mM NO_3^-
B1	1 – Fe(0)column	B1–1	100	–	–	16 h	
	2 – Bio column (NO_3^-)	B1–2	–	80	20	16 h	
System C: Partially mixed Fe(0)/biowall followed by an anaerobic biowall							Aquifer = MB–aquifer mixture TEA: Not added
C	1 – mixed zone	C1–1	20	60	20	1 w	
	2 – bio zone	C1–2	–	78	22	2 w	
System D: sequential and mixed barriers under sulfate reducing conditions							Aquifer = MB–aquifer enriched under sulphate reducing conditions TEA: 0.5 mM SO_4^{2-}
D1	1 – Bio column (SO_4^{2-})	D1–1	–	80	20	22 h	
D2	1 – Fe(0)column	D2–1	65	35	–	22 h	
	2 – Bio column (SO_4^{2-})	D2–2	–	80	20	22 h	
D4	1 – Mixed column (SO_4^{2-})	D4–1	63	24	13	20 h	
System E: sequential and mixed barriers under iron reducing conditions							Aquifer = MB–aquifer enriched under iron reducing conditions TEA: 2.5 mM Fe(III)EDTA
E1	1 – Bio column (Fe^{3+})	E1–1	–	80	20	19 h	
E2	1 – Fe(0)column	E2–1	65	35	–	21 h	
	2 – Bio column (Fe^{3+})	E2–2	–	80	20	19 h	
E4	1 – Mixed column (Fe^{3+})	E4–1	63	24	13	21 h	

TEA, terminal electron acceptor; HRT, estimated hydraulic retention time; h, hours; w, weeks; FS, filter sand.

Table 3 Organic pollutant removal efficiencies (%) in the different multibarrier columns.

Column trains	Characteristics (TEA)	ID	PCE	TCE	Benzene	Toluene	*m*-xylene
System A: Zero valent iron wall followed by a micro-aerobic biobarrier [1]							
A1	1 – Fe(0)column	A1-1	60–80	90–99	4–13	10–53	NA
	2 – Bio column (O_2)	A1-2	80–100	92–100	99–100	99–100	> 90
System B: Zero valent iron wall followed by a denitrifying biobarrier [1]							
B1	1 – Fe(0)column	B1-1	66–99	97–100	6–18	16–54	NA
	2 – Bio column (NO_3^-)	B1-2	80–93	97–100	89–100	99–100	> 90
System C: Partially mixed Fe(0)/biowall followed by anaerobic biowall [2, 1]							
C1	1 – mixed zone	C1-1	95–100 [1]	95–100 [1]	20 [2]	30 [2]	40 [2]
	2 – bio zone	C1-2	95–100 [1]	95–100 [1]	20 [2]	30 [2]	50 [2]
System D: sequential and mixed barriers under sulfate reducing conditions [1]							
D1	1 – Bio column (SO_4^{2-})	D1-1	0–11	0–15	11–44	97–100	NA
D2	1 – Fe(0)column	D2-1	55–71	88–98	4–54	0–36	NA
	2 – Bio column (SO_4^{2-})	D2-2	55–78	89–98	23–67	12–36	<25
D4	1 – Mixed column (SO_4^{2-})	D4-1	76–89	99–100	18–62	7–23	>90
System E: sequential and mixed barriers under iron reducing conditions [1]							
E1	1 – bio column (Fe^{3+})	E1-1	0–11	9–18	24–29	100	NA
E2	1 – Fe(0)column	E2-1	53–67	85–92	50–68	72–91	NA
	2 – Bio column (Fe^{3+})	E2-2	54–67	86–92	52–71	77–90	>90
E4	1 – Mixed column (Fe^{3+})	E4-1	58–73	94–97	55–64	94–98	>90

[1] Removal efficiencies determined based on the concentration in the effluent of a (sub) column and the concentration in the influent of the column train; [2] Removal efficiencies determined relative to the poisoned control; NA, Not available;

In the column trains of system D and E PCE and TCE were only removed when Fe^0 was present. With the mixed system higher removal efficiencies were obtained in comparison with the sequential set-up. Under sulfate reducing conditions the degradation of BTX was less efficient, explicable by the observed pH increases.

None of the added heavy metals were detected in the effluents of the multibarrier systems, except in the columns where Fe(III)EDTA was added as TEA (results not shown). This is in agreement with the results of other tests where a very high retention of Zn and As on the zero valent iron and the used aquifer mixture was observed (Bastiaens *et al.*, unpublished results).

CONCLUSION

The use of zero valent iron for the remediation of groundwater polluted with certain chlorinated compounds is a proven technology (Gillham 1996; O'Hannesin & Gillham, 1998, Bastiaens *et al.*, 2002a, 2002b). As it concerns an abiotic reductive removal process, not much attention has been given to the microbiology in zero valent iron barriers. However, recently it has been shown that bacteria are present in and near iron barriers and that they may play an important role in the functioning of the barriers (Scherer *et al.*, 2000; Gu *et al.*, 2002; Van Eekert *et al.*, 2003). In particular the hydrogen formed during anaerobic corrosion of the iron may select for certain bacterial groups able to use hydrogen as an electron donor.

In the first part of the study presented experiments were performed to examine the compatibility between key-components in multibarriers, namely zero valent iron, BTX-degrading bacteria, TEA and a mixed pollutant water. Zero valent iron was found to have only an indirect negative effect on BTX-degrading micro-organisms due to pH-effects. Furthermore the presence of other pollutants reduced the benzene removal efficiency. Interestingly BTX-degrading cultures were enriched from an aquifer mixture under denitrifying, sulfate and iron reducing conditions.

In the second part of the study, the enriched cultures were used as inoculum for column set ups simulating different multibarrier concepts. Both mixed and sequential multibarrier configurations were shown to be suitable for sanitation of mixed groundwater pollution. Indications were found for an improved removal in mixed systems in some cases. In all multibarrier concepts investigated, with exception of system D, chlorinated ethenes and heavy metal removal was observed when zero valent iron was present. Biodegradation of BTX was observed under aerobic, denitrifying and iron reducing conditions, but to a much lesser extent when sulfate was present as TEA.

In the third part of this work, a partially mixed multibarrier was installed on a pilot scale in a container system (5 m × 2.4 m × 2.4 m) in which an aquifer is simulated. The tested multibarrier consists of a mixed Fe^0+BIO zone followed by an anaerobic BIO-zone and a sorption zone. Iron (III) originating from the corrosion of the zero valent iron and sulfate present in the groundwater were selected as TEA. Besides monitoring of the chemical composition of the groundwater and the field parameters, *in situ* mesocosm socks and molecular techniques like PCR-DGGE are being used to monitor changes in the microbial population in the different zones.

Acknowledgements This work was funded by a VITO PhD grant (to J. Dries) and by EC project no. QLRT-2000-00163 (multibarrier).

REFERENCES

Bastiaens, L., Maesen, M., Vos, J., Kinnaer, L., Diels, L., Nuyens, D. & O'Hannesin, S. (2002a). Permeable reactive iron barrier for treatment of groundwater highly polluted with TCE. In: *Remediation of Chlorinated and Recalcitrant Compounds* (ed. by A. R. Gavaskar & A. S. C. Chen), (Proc. Third Int. Conf. Monterey, California, May 2002), Paper 2A-08. Battelle Press, Columbus, Ohio, USA..

Bastiaens, L., J. Vos, M. Maesen, L. Kinnaer, L. Diels, K. Weytingh, A. van de Velde, E. Berndsen, A. Peene & S. O'Hannesin (2002b) Feasibility studies, design and implementation of a European permeable reactive iron barrier. In: *Remediation of Chlorinated and Recalcitrant Compounds* (ed. by A. R. Gavaskar & A. S. C. Chen), Paper 2A-23. (Proc. Third Int. Conf. Monterey, California, May 2002). Battelle Press, Columbus, Ohio, USA.

Dries, J., Hendrickx, H., Bastiaens, L., Springael, D., Agathos, S. N. & Diels, L. (2002) A sequential reactive barrier concept for remediation of groundwater containing mixed pollution. In: *Remediation of Chlorinated and Recalcitrant Compounds* (ed. by A. R. Gavaskar & A. S. C. Chen), Paper2A-22. (Proc. Third Int. Conf. Monterey, California, May 2002). Battelle Press, Columbus, Ohio, USA..

Dries (2004) Development and comparison of different multifunctional permeable reactive barrier (MPRB) concepts for the treatment of groundwater contaminated by pollutant mixtures. PhD-thesis 50/2004, University Catholique de Louvain, Belgium.

Dries, J., Geuens, S., Bastiaens, L., Springael, D., Agathos, S. N. & Diels, L. (2004) Multibarrier, a technology concept for *in-situ* remediation of mixed groundwater pollution. In: *in situ and On-Site Bioremediation-2003* (ed. by V. S. Magar & M. E. Kelley) (Proc. Seventh Int. Symp., Orlando, Florida, June 2003), Paper K-04. Battelle Press, Columbus, Ohio, USA.

Gillham, R. W. (1996) *In-situ* treatment of groundwater: metal-enhanced degradation of chlorinated organic contaminants, In: *Advances in Groundwater Pollution Control and Remediation* (ed. by M. M. Aral), 249–274. Kluwer Academic Publishers, Dordrecht, The Netherlands.

Gu, B., Watson, D. B. Wu, L., Phillips, D. H. White, D. C. & Zhou, J. (2002) Microbiological characteristics in a zero valent iron reactive barrier. *Environ. Monitoring and Assessment* **77**, 293–309.

O'Hannesin, S. F & Gillham, R. W. (1998) Long-term performance of an *in-situ* "Iron Wall" for remediation of VOCs. *Ground Water* **36**, 164–170.

Scherer, M. E., Richter, S., Valentine, R. L. & Alvarez. P. J. J. (2000) Chemistry and microbiology of permeable reactive barriers for *in-situ* groundwater clean up. *Critical Rev. Environ. Sci. & Technol.* **30**(3), 363–411.

Van Eekert, M., Draaisma, C., Diels, L. & Schraa., G. (2003) Treatment of multiple pollution in a multibarrier system. In: *Contaminated Soil (CONSOIL 2003)* (Conf. Proc. Eighth Int. FZK/TNO Conf., Ghent, Belgium, 12–16 May 2003), 1666–1674.

Permeable Reactive Barriers (Proceedings of the International Symposium held at Belfast, Northern Ireland, March 2004). IAHS Publ. 298, 2005.

The development and use of permeable reactive barrier technologies and potential future applications in the UK

B. D. BONE[1], R. C. HARRIS[1], J. W. N. SMITH[1], G. A. BOSHOFF[2], R. M. KALIN[3], R. THURGOOD[4] & P. MORGAN[5]

1 *Environment Agency Science Group, Olton Court, 10 Warwick Road, Olton, Solihull, West Midlands B92 7HX, UK.*
brian.bone@environment-agency.gov.uk

2 *Mouchel Parkman, Parkman House, Lloyd Drive, Ellesmere Port, South Wirral CH65 9HQ, UK*

3 *Environmental Engineering Research Centre, School of Civil Engineering, The Queen's University Belfast, Belfast BT9 5AG, UK.*

4 *CEL International Ltd, 256 Foleshill Road, Coventry CV6 5AB, UK.*

5 *ESI, Priory House, Priory Road, Shrewsbury SY1 1RU, UK.*

Abstract A regulatory approach has been developed by the Environment Agency (the Agency), and linked to good practice guidance, to encourage the appropriate use of permeable reactive barriers (PRBs) in England and Wales. Innovative approaches to PRB technologies are being developed in the UK and taken forward to full-scale implementation. These developments, although small in number, utilize a variety of reactive media to address a range of contaminants, including mixed contaminant plumes. Potential for further development of PRB technologies exists, in particular to reduce the impact of diffuse groundwater pollution. In this context, PRB technologies may help to reduce the impact of groundwater pollution on aquatic ecosystems, even at river-reach to catchment-scale.

Key words groundwater; permeable reactive barrier; pollution; PRB; remediation

INTRODUCTION

The full-scale implementation of permeable reactive barriers (PRBs) has a track record of over a decade for the treatment of groundwater contaminated with chlorinated hydrocarbons using zero valent iron (Fe^0) (O'Hannesin, 2003). This technology can utilize a number of abiotic and/or biotic reactions to reduce the contaminant mass or mobility, and has been taken up extensively, with over 110 pilot scale and field scale PRBs constructed in North America, Europe and the UK (Smith *et al.*, 2003; O'Hannesin, 2003).

The UK market has been slow to accept PRBs as a remediation strategy. However, a number of initiatives have been taken to support the appropriate use of PRB technologies, including the introduction of good practice guidance (Environment Agency, 2002) and a regulatory enforcement position in England and Wales. Together, these actions will enhance confidence in the use of well-designed, constructed and operated PRBs.

Regulatory approach in England and Wales

The Agency is responsible for regulating remediation activities for contaminated soil and groundwater in England and Wales. The Agency aims to ensure that the selected remediation solution removes unacceptable risks to the environment and human health effectively, and that the remedial activities do not cause harm or pollution. The use of cost-effective and sustainable remediation technologies, including PRBs, is encouraged, and the Agency intends to regulate proportionately to the risks associated with the remedial activities.

Under both European Union (Waste Framework Directive, 75/442/EEC, as amended by 91/156/EEC) and national (Environmental Protection Act, 1990) legislation, contaminated soil and groundwater are usually considered to be controlled wastes and the treatment of such a material is therefore a waste recovery or disposal activity. Consequently, unless a specific exclusion or exemption to the regulations applies, remediation activities require authorization under the relevant waste legislation.

The Agency considers that the application of a full waste management licence for the installation of a PRB would be disproportionate in terms of the potential risk to the environment caused by its installation, and would act as a disincentive to the use of the technology. Since the use of effective and sustainable remedial solutions is to be encouraged, the Agency has developed an enforcement position that essentially states that enforcement action will not be taken against a PRB operator for failing to hold a waste management licence. This is on condition that the Agency is notified of their intention to install a PRB, good practice is used in its design, construction, operation and monitoring, and the remedial activity does not cause harm or pollution. Where good practice is not adopted, or pollution or harm caused, the Agency retains the right to take enforcement action.

To support the enforcement position, the Agency published good practice guidance in 2002 (Environment Agency, 2002) that sets out advice on:

(a) permitting of PRBs;
(b) screening procedures to assess the viability of a PRB;
(c) PRB design, construction and operation;
(d) monitoring the performance of a PRB; and
(e) PRB decommissioning.

The guidance follows a tiered framework that is designed to ensure that resources are focused at collecting the necessary data to demonstrate the likely effectiveness of a PRB and on which to base its design. In the longer term, verification of treatment performance through monitoring will be necessary and data may need to be collected to design an appropriate decommissioning strategy.

A key requirement of good practice is the development of a conceptual model that describes the hydrogeological, biochemical and geochemical characteristics of a site and how the PRB will interact with the contaminant flux and groundwater flow regime. Key potential failure mechanisms or factors that may need to be addressed include mineral precipitation (Borden *et al.*, 1997; Kamolpornwijit *et al.*, 2003; Liang *et al.*, 2003; Morrison, 2003; Phillips *et al.*, 2003), biofouling (Weathers *et al.*, 1997; Gu *et al.*, 1999), gas clogging (Mackenzie *et al.*, 1999), competition between contaminants or between contaminants and other dissolved species (Johnson *et al.*,

1998; Tratnyek *et al.*, 2001; Schäfer *et al.*, 2002; Su & Puls, 2003), desorption/remobilization (Farrell *et al.*, 1999; Krumhansl *et al.*, 2001; Benner *et al.*, 2002), temperature (Benner *et al.*, 2002) and pH changes (Chen *et al.*, 2001). There may also be potential for the generation of polluting substances within the barrier as a consequence of secondary reactions not considered as part of the PRB design for its target pollutants (Cheng *et al.*, 1997). In addition, temporal changes in groundwater flow direction may result in migration of the plume away from or around the PRB.

Monitoring also needs to be undertaken to demonstrate whether the PRB is effective as a remedial solution and that it is not causing additional pollution or harm. Consequently, monitoring may be required for a long period of time: before construction and for a period of years to verify the continued performance of the PRB and to determine when treatment can cease. Additional monitoring is likely for a period following cessation and/or decommissioning.

The operator is also required to have a decommissioning plan that can be implemented once the remedial objectives have been achieved. This will take into account whether the PRB should be removed, left in place or other works undertaken, and is required since:

(a) a future change in site conditions (e.g. geochemical environment) may result in remobilization of contaminants from the reactive material (e.g. by subsequent desorption of sorbed substances or remobilization of precipitated substances);
(b) the permeability of the PRB may change, due to clogging or fouling processes with the potential for an unacceptable rise in groundwater level up-gradient of the barrier; and
(c) the deterioration of the reactive media may itself give rise to pollution.

As well as regulating remediation activities, the Agency has a duty to contribute to the goal of achieving sustainable development, which it does through exercising its regulatory powers and by encouraging others to select more sustainable remedial options where they are practicable, effective, durable and reasonable in terms of cost. The successful uptake of innovative, sustainable technologies in the UK will depend on partnership between researchers, remediation industry, problem holders and regulators. A number of organizations have been established in the UK to promote partnerships and/or disseminate knowledge to a wide audience. Such initiatives include CL:AIRE (Contaminated land: applications in the real environment), FIRSTFARADAY (Faraday partnership for innovative remediation science and technology) and, for PRB technologies, PRB-Net (www.prb-net.org).

Innovative technologies developed in the UK

Despite the relatively small number of PRBs so far installed, the UK can boast a wide variety of approaches that have been tried (Smith *et al.*, 2003), including the first Fe^0 PRB installed in Europe, in 1995 (Beck *et al.*, 2001). Active research into new PRB technologies and evaluation of the performance of installed systems is being carried out at a number of UK institutions, and research has been successfully transferred to full-scale field evaluation at a number of sites, including the three discussed below.

Acid mine drainage

Amos & Younger (2003) describe the first large-scale continuous PRB in the UK for treating leachate from mine spoil at Shilbottle Colliery, Northumberland. The leachate is characterized by low pH (~3.5–4.5) and high concentrations of iron (to ~1000 mg l^{-1}), aluminium (to ~700 mg l^{-1}) and manganese (to ~300 mg l^{-1}). Laboratory trials were used to identify reactive media able to reduce acidity, generate alkalinity and promote biologically mediated sulphate reduction and precipitation of metal sulphides. The reactive media used comprise a mixture of limestone gravel (50%), green waste compost (25%) and animal manure (25%). The barrier was installed in 2002 and is currently being extensively monitored (Younger & Moustafa, 2005) to assess the system performance and to enhance the future design of similar systems.

Mixed contaminant plume from former gasworks site

Kalin (2004) presented progress on the installation of a sequenced reactive barrier (SEREBAR) at a former coal gasification site in southwest England. The SEREBAR project is a collaborative initiative to assess the potential for a sequential biobarrier to address mixed contaminant plumes characteristic of gasworks sites, including polycyclic aromatic hydrocarbons (PAHs), ammonium, cyanide, phenols and BTEX. This project combines both laboratory scale experiments to evaluate the treatability of the contaminants under aerobic and anaerobic bioactive conditions and a full-scale technology demonstration (supported by CL:AIRE), to evaluate sequential barriers using Fe^0, air-sparged sand boxes and granular activated carbon components.

Based on the laboratory experiments, a full-scale PRB was successfully commissioned in April 2004. Its performance will be monitored to assess the contaminant removal efficiency, study the microbial communities that mediate degradation, identify the optimum conditions to maintain within the PRB, and evaluate the risk of biofouling to the long-term performance of the PRB. This is intended to lead to the design of modelling tools for the design and predictive evaluation of the performance of sequenced reactive barriers.

Carbon disulphide

A novel method for treating groundwater contaminated by dissolved phase CS_2 was developed at the Queen's University, Belfast (QUB) (Kalin & Myles, 2003) and implemented as part of a remedial strategy at a former chemical works in northwest England. The site was to be redeveloped for light industrial use and works were required to manage the risks associated with the contaminants, primarily carbon disulphide. The remedial strategy included:

(a) removal of CS_2-contaminated soils by bentonite slurry stabilization, excavation and disposal off-site to approved landfills. Stabilization was essential to permit safe handling and transport of the material, which is highly combustible; and
(b) construction of a "funnel & gate" PRB to reduce the contaminant flux in shallow groundwater crossing the site boundary.

The use of Fe^0 for the treatment of some chlorinated hydrocarbons is a proven technology (Gillham, 1996). However, its use for CS_2 has never before been demonstrated. Experimental work at QUB indicated that Fe^0 may be effective for the degradation of dissolved phase CS_2 via carbonyl sulphide to iron sulphide, oxyhydroxide and carbonate precipitates (Kalin & Myles, 2003; Kalin *et al.*, 2005). Following laboratory studies, utilizing real-time and accelerated column studies, a double-gate funnel-and-gate system was designed using a conservative abiotic half-life for CS_2 in the presence of Fe^0 of 4 hours. Each gate comprises two parallel Fe^0-filled downflow–upflow vertical reactors (Fig. 1), designed for ease of maintenance and cost-effective decommissioning. The system was designed to treat a hydraulic flux of 75 m^3 day^{-1} at an average CS_2 concentration of 10 mg l^{-1} and short-term peak of 100 mg l^{-1}. The gates were installed in May 2004, and this site has been proposed as a CL:AIRE technology demonstration project to make the case study of this innovative PRB technology available to a wider audience.

Fig. 1 Installation of one of two gates showing two vertical reactors (two pairs in parallel per gate).

Future application of PRBs to address diffuse pollution

The Water Framework Directive (2000/60/EC–WFD) will rationalize existing water legislation and introduce an integrated approach to water management in the European Union based on the concepts of river basin planning and management.
The major aims of the Directive, as set out in Article 1, are:

(a) to prevent further deterioration and protect and enhance the status of aquatic ecosystems and associated wetlands;
(b) to promote the sustainable consumption of water;
(c) to reduce pollution of waters from priority substances;

(d) to prevent the deterioration in the status and to progressively reduce pollution of groundwaters; and
(e) to contribute to mitigating the effects of floods and droughts.

The overall requirement of the WFD is to achieve "good ecological and good chemical status" in all water bodies by 2015 unless there are grounds for derogation. There is also a general "no deterioration" provision to prevent deterioration in the status of water bodies. Member states will have to:

(a) characterize river basin catchments;
(b) manage "pressures" within them; and
(c) reduce "impacts" on ecosystems considered in a whole river basin context within a "Programme of Measures" to ensure water bodies reach the environmental objectives.

Improvements to aquatic (eco)systems can only be realised by changing land management and/or remediating historic pollution. Pollution sources that may have a major impact at a catchment-scale are indicated in Table 1.

Table 1 Catchment scale pollution sources and solutions.

Pollution source	Potential solutions
Agricultural diffuse pollution (e.g. nitrate, phosphate, pesticides)	Radical land use change required, pathway interception for historic contamination
Historical mining activity (point and diffuse sources)	Complex, multi-source sites may require a possible mixture of source control, pathway interception and targeted receptor protection
Large scale urban pollution (including "mega-sites")	Very difficult to deal with source but combination of pathway interception and targeted receptor protection may be possible
Point sources (few likely)	Consider under site-scale hierarchy of source treatment, pathway interception and receptor protection

In the UK, diffuse pollution sources are considered to represent the most significant challenges to achieving WFD targets. Significant diffuse sources include agriculture (nitrate, phosphate, pesticides) and mining (acidity, metals, sulphate). Although improved land use management can reduce future releases of pollutants, historic contamination of soil and groundwater may prove impractical to treat at source.

It is likely that passive, low-impact technologies, such as monitored or enhanced natural attenuation and PRBs, will become increasingly important in the management of river catchments as part of a programme of works to reduce impacts. This may be the case particularly where they can be incorporated into other structures, such as land drains, flood defences and urban drainage, and utilize low-cost, readily available materials. PRB technologies are already being studied at laboratory and field scale to address significant diffuse pollution sources from agriculture (Robertson *et al.*, 2000; Schipper & Vojvodić-Vuković, 2000; Alowitz & Scherer, 2002; Moon *et al.*, 2004) and mines (Amos & Younger, 2003; Waybrant *et al.*, 2002). PRB technologies show great potential for either dealing with historic diffuse pollution or to be integrated with other pollution prevention measures to improve land management.

CONCLUSIONS

There has been a slow but steady increase in the use of PRB technologies in the UK (Smith *et al.*, 2003), supported by regulatory acceptance, innovative research carried out at a number of academic institutions, and partnership with a number of organizations to promote knowledge transfer, including CL:AIRE and FIRSTFARADAY. Ten years after the first commercial application of a PRB in the UK, there is now a significant global track record for the treatment of some chlorinated hydrocarbons using Fe^0 (e.g. Liang *et al.*, 2001; Kamolpornwijit *et al.*, 2003; Wilkin & Puls, 2003; USEPA, 2004; Warner *et al.*, 2005). However, the use of other contaminant–reactive media combinations has a less well established track record, but there is a growing body of research into the use of other media in isolation, in sequence or combined (e.g. Amos & Younger, 2003; Bastiaens *et al.*, 2005; Kalin, 2005) to treat a wider range of contaminant plumes. This is a key research area if the challenges posed by mixed contaminant plumes and diffuse pollution sources are to be addressed by low cost, low impact yet effective technologies. PRBs represent a potentially sustainable technology for the treatment of contaminated groundwater, but the cost and impact of maintenance and decommissioning, if required, need to be taken into account for a PRB to be considered sustainable at any particular site.

Acknowledgements The authors would like to acknowledge the support of the Environment Agency, CL:AIRE and FIRSTFARADAY, and to the research councils Engineering and Physical Sciences Research Council (EPSERC) and Biotechnology and Biological Sciences Research Council (BBSRC). The views expressed in this paper are those of the authors and do not necessarily reflect the views or policies of their respective organisations.

REFERENCES

Alowitz, M. J. & Scherer, M. M. (2002) Kinetics of nitrate, nitrite and Cr(VI) reduction by iron metal. *Environ. Sci. Technol.* **36**, 299–306.

Amos, P. W. & Younger, P. L. (2003) Substrate characterisation for a subsurface reactive barrier to treat colliery spoil leachate. *Water Res.* **37**, 108–120.

Bastiaens, L., Dries, J., Vos, J., Simons, Q., De Smet, M. & Diels, L. (2005) Comparison of different multibarrier concepts designed for treatment of groundwater containing mixed pollutants. In: *Permeable Reactive Barriers* (ed. by G. A. Boshoff & B. D. Bone) (Proc. Int. Symp., Belfast, Northern Ireland, March 2004), 45–51. IAHS Publ. 298. IAHS Press, Wallingford, UK (this volume).

Beck, P., Harries, N. & Sweeney, R. (2001) Design, installation and performance assessment of a zero valent iron permeable reactive barrier in Monkstown, Northern Ireland. *CL:AIRE Technology Demonstration Report TDP3*.

Benner, S. G., Blowes, D. W., Ptacek, C. J. & Mayer, K. U. (2002) Rates of sulfate reduction and metal sulfide precipitation in a permeable reactive barrier. *Appl. Geochem.* **17**, 301–320.

Borden, R. C., Goin, R. T. & Kao, C. M. (1997) Control of BTEX migration using a biologically enhanced permeable barrier. *Ground Water Monitoring Remed.* **17**, 70–80.

Chen, J. L., AlAbed, S. R., Ryan, J. A. & Li, Z. (2001) Effects of pH on dechlorination of trichloroethylene by zero-valent iron. *J. Hazard. Mat.* **B83**, 243–254.

Cheng, F., Muftikian, R., Fernando, Q. & Korte, N. (1997) Reduction of nitrate to ammonia by zero valent iron. *Chemosphere* **35**, 2689–2695.

Environment Agency (2002) Guidance on the use of permeable reactive barriers for remediating contaminated groundwater. *National Groundwater and Contaminated Land Centre Report NC/01/51, Environment Agency, Solihull, UK.*

Farrell, J., Bostick, W. D., Jatabek, R. J. & Fiedor, J. N. (1999) Uranium removal from ground water using zero valent iron media. *Ground Water* **37**, 618–624.

Gillham, R. W. (1996) *In-situ* treatment of groundwater: metal-enhanced degradation of chlorinated organic contaminants. In: *Advances in Groundwater Pollution Control and Remediation* (ed. by M. M. Aral), 249–274. Kluwer Academic Publishers, Dordrecht, The Netherlands.

Gu, B, Phelps, T. J., Liang, L., Dickey, M. J., Kinsall, B. L. & Jacobs, G. K. (1999) Biogeochemical dynamics in zero-valent iron columns: implications for permeable reactive barriers. *Environ. Sci. Technol.* **33**, 2170–2177.

Johnson, T. L., Fish, W., Gorby, Y. A. & Tratnyek, P. G. (1998) Degradation of carbon tetrachloride by iron metal: complexation effects on the oxide surface. *J. Contam. Hydrol.* **29**, 379–398.

Kalin, R. M. (2005) *In-situ* bioremediation of cyanide, PAHs and organic compounds using engineered sequenced reactive barrier (SEREBAR) techniques: a progress report 0 to 14 months. In: *CL:AIRE Annual Project Conference*, (Unversity College, London, 20 April, 2004).

Kalin, R. M. & Myles, P. (2003) A method for the treatment of carbon disulphide in soil and groundwater. *International Patent Application No. PCT/GB2003/002012.*

Kalin, R. M., Myles, P. & McGeogh, K. (2005) Carbon disulphide sources, risks and treatment: an emerging contaminant of environmental concern? In: *Permeable Reactive Barriers* (ed. by G. A. Boshoff & B. D. Bone) (Proc. Int. Symp., Belfast, Northern Ireland, March 2004), 82–92. IAHS Publ. 298. IAHS Press, Wallingford, UK (this volume).

Kamolpornwijit, W., Liang, L., West, O. R., Moline, G. R. & Sullivan, A. B. (2003) Preferential flow path development and its influence on long-term PRB performance: column study. *J. Contam. Hydrol.* **66**, 161–178.

Krumhansl, J. L., Brady, P. V. & Anderson, H. L. (2001) Reactive barriers for ^{137}Cs retention. *J. Contam. Hydrol.* **47**, 233–240.

Liang, L., Korte, N. E., Moline, G. R. & West, O. R. (2001) *Long-term Monitoring of Permeable Reactive Barriers*, ORNL/TM-2001/1. Oak Ridge National Laboratory, Oak Ridge, Tennessee, USA.

Liang, L., Sullivan, A. B., West, O. R., Kamolpornwijit, W. & Moline, G. R. (2003) Predicting the precipitation of mineral phases in permeable reactive barriers. *Environ. Eng. Sci.* **20**(6), 635–653.

Mackenzie, P. D., Horney, D. P. & Sivavec, T. M. (1999) Mineral precipitation and porosity losses in granular iron columns. *J. Hazard. Mat.* **68**(1-2), 1–17.

Moon, H. S., Ahn, K. -H., Lee, S., Nam, K. & Kim, J. Y. (2004) Use of autotrophic sulfur-oxidizers to remove nitrate from bank filtrate in a permeable reactive barrier system. *Environ. Poll.* **129**, 499–507.

Morrison, S. J. (2003) Performance evaluation of a permeable reactive barrier using reaction products as tracers. *Environ. Sci. Technol.* **37**, 2302–2309.

O'Hannesin, S. F. (2003) 10 years of north American experience in granular iron PRB technology for VOC groundwater remediation. In*: Proceedings of the Eighth International FZK/TNO Conference on Contaminated Soil* (Ghent. Belgium). CD only.

Phillips, D. H., Watson, D. B., Roh, Y. & Gu, B. (2003) Mineralogical characteristics and transformations during long-term operation of a zerovalent iron reactive barrier. *J. Environ. Qual.* **32** (6), 2033–2045.

Robertson, W. D., Blowes, D. W., Ptacek, C. J. & Cherry, J. A. (2000) Long-term performance of *in situ* reactive barriers for nitrate remediation. *Ground Water* **38**, 689–695.

Schäfer, D., Kober, R., Ebert, M. & Dahmke, A. (2002) Effects of interspecies competition on the design of permeable reactive iron barriers. In: *Proceedings of Geochemical Processes 2002* (4–7 March, Bremen, Germany).

Schipper, L. A. & Vojvodić-Vuković, M. (2000) Nitrate removal from groundwater and denitrification rates in a porous treatment wall amended with sawdust. *Ecol. Eng.* **14**, 269–278.

Smith, J. W. N., Boshoff, G. & Bone, B. D. (2003) Good practice guidance on permeable reactive barriers for remediating polluted groundwater, and a review of their use in the UK. *Land Contam. and Reclam.* **11**, 411–418.

Su, C. & Puls, R. W. (2003) *In situ* remediation of arsenic in simulated groundwater using zero valent iron: laboratory column tests on combined effects of phosphate and silicate. *Environ. Sci. Technol.* **37**, 2582–2587.

Tratnyek, P. G., Scherer, M. M., Deng, B. & Hu, S. (2001) Effects of natural organic matter, anthropogenic surfactants, and model quinines on the reduction of contaminants by zero-valent iron. *Water Research* **35**, 4435–4443.

USEPA (2004) *Evaluation of Permeable Reactive Barrier Performance*. Prepared for Federal Remediation Technology Roundtables by the tri-agency Permeable Barrier Initiative, US Department of Defense, US Department of Energy, US Environmental Protection Agency, Interstate Technology and Regulatory Cooperation, 2002, EPA 542-R-04-004.

Warner, S. D., Longino, B. L., Zhang, M., Bennett, P., Szerdy, F. S. & Hamilton, L. A. (2005) The first commercial permeable reactive barrier composed of granular iron—hydraulic and chemical performance at 10 years of operation. In: *Permeable Reactive Barriers* (ed. by G. A. Boshoff & B. D. Bone) (Proc. Int. Symp., Belfast, Northern Ireland, March 2004), 32–42. IAHS Publ. 298. IAHS Press, Wallingford, UK (this volume).

Waybrant, K. R., Ptacek, C. J. & Blowes, D. W. (2002) Treatment of mine drainage using permeable reactive barriers: column experiments. *Environ. Sci. Technol.* **36**, 1349–1356.

Weathers, L. J., Parkin, G. F. & Alvarez, P. J. (1997) Utilization of cathodic hydrogen as electron donor for chloroform cometabolism by a mixed methanogenic culture. *Environ. Sci. Technol.* **31**, 880–885.

Wilkin, R. & Puls, R. (2003) Capstone report on the application, monitoring, and performance of permeable reactive barriers for ground-water remediation. *EPA 600-R-03-045, US EPA, Cincinnati, Ohio, USA.*

Younger, P. L. & Moustafa, M. (2005) Remediation of acidic colliery spoil leachate in a hybrid passive treatment system comprising a permeable reactive barrier, ponds and a reedbed (Shilbottle, Northumberland, UK). In: *Permeable Reactive Barriers* (ed. by G. A. Boshoff & B. D. Bone) (Proc. Int. Symp., Belfast, Northern Ireland, March 2004), 117–122. IAHS Publ. 298. IAHS Press, Wallingford, UK (this volume).

Permeable Reactive Barriers (Proceedings of the International Symposium held at Belfast, Northern Ireland, March 2004). IAHS Publ. 298, 2005.

Cost-effectiveness of *in situ* treatment of multiple contaminant mixtures in groundwater by means of sequenced units of zero valent iron and granular activated carbon

MICHAEL FINKEL & PETER BAYER
Centre for Applied Geoscience, University of Tuebingen, Sigwartstrasse 10, D-72076 Tuebingen, Germany
michael.finkel@uni-tuebingen.de

Abstract This modelling study reveals that the sequential use of zero valent iron (ZVI) and granular activated carbon (GAC) can potentially be used for *in situ* treatment of groundwater with multiple contaminant mixtures, but crucial design rules should be followed, since the performance of the combined system depends on the interplay of both reactor units. As an example a mixture of aliphatic chlorinated hydrocarbons (PCE, TCE, and *cis*-DCE) and one aromatic chlorinated hydrocarbon (monochlorobenzene) is considered. The results show that the concentrations of less sorbing compounds such as *cis*-DCE have to be reduced in the ZVI reactor far beyond the desired maximum contaminant level in order to restrict concentration build-ups in the GAC reactor, which occur due to competitive sorption with contaminants withheld in the GAC. Results from reactor modelling studies are used to economically analyse feasible reactor designs. Cost-optimal ZVI-GAC combinations are derived by a general approach based on unit cost ratios.

Key words competitive sorption; dechlorination; economic assessment; granular activated carbon; permeable reactive barriers; pore surface diffusion model; zero valent iron

MOTIVATION

Only in some instances are contaminant plumes characterized by the occurrence of one unique type of pollutant. In the case of a mixture of contaminants, especially if they span a broad range of physico-chemical properties, the treatment of groundwater can be a challenge. Selection of granular activated carbon (GAC), which is suited for nearly all organic compounds, might be a proper choice if a robust system applicable for complex VOC mixtures is desired. Alternatively, recent developments support combinations of a number of specified technologies such as sequential treatment technologies with zero valent iron (ZVI), bioreactors and granular activated carbon (GAC; Fiorenza *et al.*, 2000; Bell *et al.*, 2003). However, as groundwater passing through reactive media is chemically altered, treatment performance of any sequenced media may be strongly influenced by preceding treatment units. Hence, one has to be careful in addressing combined clean-up technologies as straightforward combinations of the single technologies.

The presented study discusses design considerations for the sequential use of ZVI and GAC for sanitation of groundwater contaminated by a mixture of aliphatic and aromatic chlorinated hydrocarbons. The coupling of a ZVI reactor and a GAC reactor is a promising combination for the clean-up of groundwater contaminated by organics being either suited for sorptive removal or abiotic degradation (Köber *et al.*, 2001; Kouznetsova *et al.*, 2003). Here a four-compound system consisting of monochlorobenzene (MCB), perchloroethylene (PCE), trichloroethylene (TCE), and *cis*-dichloroethylene (*cis*-DCE) is considered. MCB is assumed to be not subject to degradation within the ZVI unit and thus GAC has to be coupled to the outlet of the ZVI-reactor to completely clean up the groundwater. We use a modelling framework to simulate relevant degradation and sorption processes, as well as to glance at contaminant concentrations at the interface between both reactors. The modelling results for a hypothetical scenario serve as the basis for an exemplary economic calculation in order to reveal crucial design criterions of the combined system.

MODELLING TOOLSET

The ZVI reactor model simulates the pseudo-first order degradation mechanism which is typical for low to medium concentrations (<200 μMol according to Wüst *et al.*, 1999) of contaminants such as chlorinated ethenes. The comprehensive reaction framework describing the dechlorination from perchloroethylene (PCE) to trichloroethylene (TCE), dichloroethylenes (DCE), vinyl chloride (VC) and finally ethane is given by Arnold & Roberts (2000). Within this study, a reduced model is used which focuses on the main hazardous compounds PCE, TCE and *cis*-DCE (Bayer, 2003). For the presented generic study, VC and further DCE isomers are not considered but may play an important role in other cases.

The outflow of the ZVI reactor represents the inflow of the GAC reactor, which is modelled by the pore surface diffusion model described by Crittenden *et al.* (1986). It incorporates the main kinetic processes of single and multiple contaminant sorption on GAC. Competitive sorption, which leads to displacement of less sorbing compounds in favour of better sorbing compounds, is simulated based on the ideal adsorbed solution theory (IAST).

The model toolset is applied to a hypothetic example. The above-mentioned four-compound system is to be treated to reach maximum contaminant levels (MCL = 10 μg l^{-1}) at the outlet of the combined reactor.

Influent concentrations of PCE, TCE, and monochlorobenzene are set to C_i = 100 μg l^{-1}. Three cases are distinguished with *cis*-DCE concentrations of $C_{cis\text{-}DCE}$ = 0, 1, 10 and 100 μg l^{-1} (Fig. 1). These are selected in order to highlight the relevance of

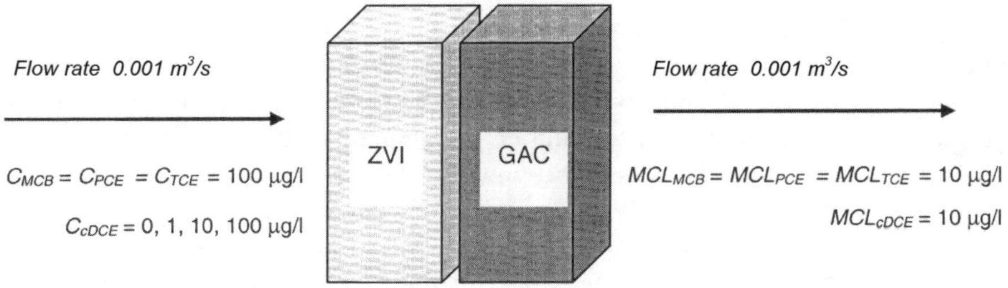

Fig. 1 Configuration of confined reactor with considered contaminant inflow concentrations (C_{PCE}, C_{TCE}, C_{DCE}, C_{MCB}) and target concentrations (MCL_{PCE}, MCL_{TCE}, MCL_{DCE}, MCL_{MCB}).

Table 1 Degradation rate constants, production rates and sorption parameters (Thiem *et al.*, 2000) used in the study.

Contaminant		Rate constant $k_i(h^{-1})$	Freundlich isotherm parameters K_F (µmol g^{-1})(l µmol^{-1})^{nF}	n_F
Monochlorobenzene	MCB		341	0.4
Perchlorothylene	PCE	0.3527	310	0.52
Trichloroethylene	TCE	0.5694	111	0.59
cis-Dichloroethylene	*cis*-DCE	0.0432	38	0.53
Production rates				
PCE → TCE	κPCE,TCE	0.137		
PCE → *cis*-DCE	κPCE,cDCE	0.015		
TCE → *cis*-DCE	κTCE,cDCE	0.262		

this compound, which occurs as a by-product (breakdown product) in the ZVI reactor and has a relatively low sorptivity. The considered degradation and production rates for treatment with ZVI as well as the Freundlich parameters describing the sorption on GAC are listed in Table 1. The volumetric flow rate of the groundwater to be treated is set constant at 0.001 m^3 s^{-1} over a planning period of 10 years.

RESULTS

Feasible designs of the sequenced ZVI-GAC reactor are determined for a range of ZVI volumes, V_{ZVI}, from 0 to 750 m^3. If V_{ZVI} is insufficient to properly reduce the concentrations of the chlorinated ethenes, GAC can be used to collect the surplus. This means that for every V_{ZVI} a GAC volume V_{GAC} exists, which is necessary to reach the MCLs at the combined reactor outlet. At first glance this aspect seems to be an advantage of the combined system.

Figure 2(a)–(d) depicts the V_{ZVI}–V_{GAC} trade-offs for each contaminant. The envelope of all contaminant-specific trade-off curves defines the required V_{GAC} for any given V_{ZVI}. Figure 2(a) reveals that the relatively low sorptivity of *cis*-DCE controls the V_{GAC} over a wide range of given V_{ZVI}. It is evident that due to the high production rate of *cis*-DCE this appears even if this contaminant is not in the influent water. Raising the influent concentration of *cis*-DCE beyond 1 µg l^{-1} (Fig. 2(c)–(d)) leads to a significant rise of V_{GAC} if the ZVI unit ahead insufficiently decreases the concentration of *cis*-DCE before it reaches the GAC reactor.

One central aspect for the design of the combined reactor is illustrated for the case with no *cis*-DCE in the groundwater to be treated. Figures 3(a–d) show concentration breakthrough curves within the operation time of T = 10 years and for a constant GAC volume of V_{GAC} = 1 m^3. Increasing V_{ZVI} from 0 m^3 to 35 m^3 results in less pronounced and delayed breakthroughs of the higher chlorinated PCE and TCE. Contaminant *cis*-DCE, produced in the ZVI unit, surmounts the given MCL of 10 µg l^{-1} after approximately one year. A significant further rise of V_{ZVI} to 178 m^3 can mitigate the *cis*-DCE concentration at the ZVI-outlet to MCL, but due to competitive sorption in the GAC reactor the attained degradation is still insufficient. Contaminant *cis*-DCE is desorbed in the GAC due to the occurrence of the better sorbing MCB and consequently shows concentrations beyond MCL at the GAC outlet after nearly 2 years. Though in

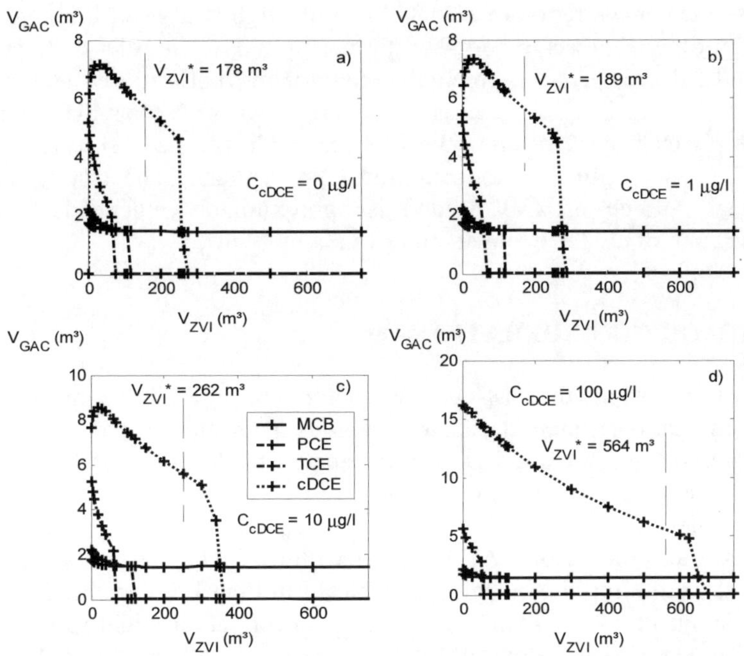

Fig. 2 Required Volume of GAC (V_{GAC}) as a function of ZVI volume (V_{ZVI}) and contaminant influent concentrations ($C_{MCB} = C_{PCE} = C_{TCE} = 100$ µg l^{-1}, $C_{cDCE} = 0, 1, 10, 100$ µg l^{-1}) for an operation time of 10 years and a constant flow rate of 0.001 m^3 s^{-1}. Vertical dashed lines indicate threshold values, V_{ZVI}^*, for which all contaminant MCLs are met at the ZVI outlet.

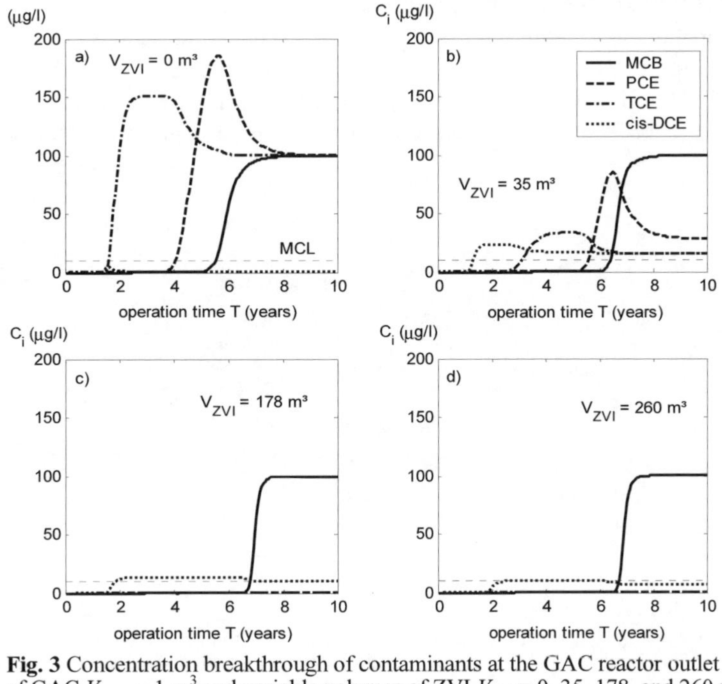

Fig. 3 Concentration breakthrough of contaminants at the GAC reactor outlet for volume of GAC $V_{GAC} = 1$ m^3 and variable volumes of ZVI $V_{ZVI} = 0, 35, 178$, and 260 m^3.

this case the concentration of *cis*-DCE is only slightly above MCL over the rest of the operation time, V_{ZVI} has to be raised further to 260 m^3 in order to meet the MCL. The threshold values, V_{ZVI}^*, for which all contaminant MCLs are reached at the ZVI outlet are given in Fig. 2 for all scenarios considered. It is obvious that independent of the influent *cis*-DCE concentration the ZVI reactor has to be overdesigned to ensure that *cis*-DCE is not critical to the required GAC volume. The results indicate that the additional volume of ZVI needed is approximately independent of the influent concentration of *cis*-DCE and amounts to about 80 m^3.

ECONOMIC CONSIDERATIONS

In order to determine the cost-optimal reactor configuration, i.e. the lowest price $V_{ZVI} - V_{GAC}$ combination, a main decision variable is the desired lifetime of the GAC reactor. The required GAC volume has been calculated for lifetimes of 1, 2.5, 5 and 10 years (number of fillings: 10, 4, 2 or 1, respectively). Figure 4 shows the corresponding $V_{ZVI} - V_{GAC}$ trade-offs (envelopes subsuming the contaminant-specific design curves). The resulting total volume of GAC required within a planning period of 10 years is not exactly the same for all strategies as depicted in Fig. 5. Lower lifetimes mean a minor accumulation of *cis*-DCE in the GAC, which effects a slightly lower GAC volume V_{GAC} if given V_{ZVI} is below 100 m^3. For V_{ZVI} > 100 m^3, the reduction of V_{GAC} is attributed to the marginal flattening of the concentration breakthrough curves with raised GAC life.

Discounting future costs gives net present values accounting for interest earnings. Assuming a discount rate of 5%, expenditures decrease especially for high number of refills, where payments occur in later stages of the planning period. We express these savings through V_{GAC} equivalents reflecting the required total V_{GAC} minus the relative discount (Fig. 6). The ideal combined system is then given by the relative material costs for ZVI and GAC. Due to the characteristic form of the $V_{ZVI} - V_{GAC}$ (equivalent) curves, a threshold value u of the unit costs ratio r = (price of ZVI per m^3)/(price of

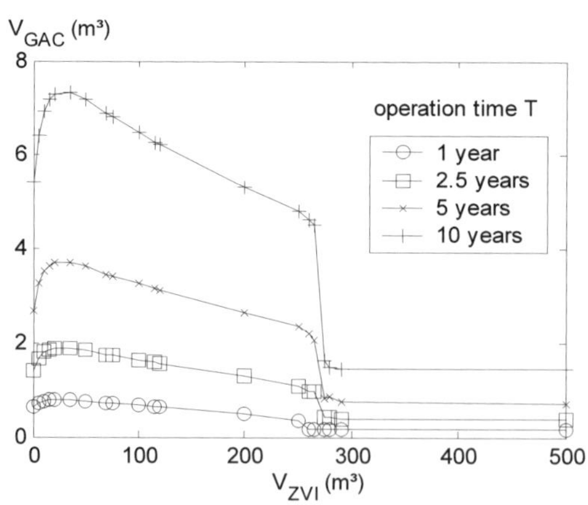

Fig. 4 Required volumes of GAC and ZVI to comply with target concentrations of all contaminants at the combined reactor outlet for four different operation times T, i.e. life times of GAC reactor (1, 2.5, 5, 10 years).

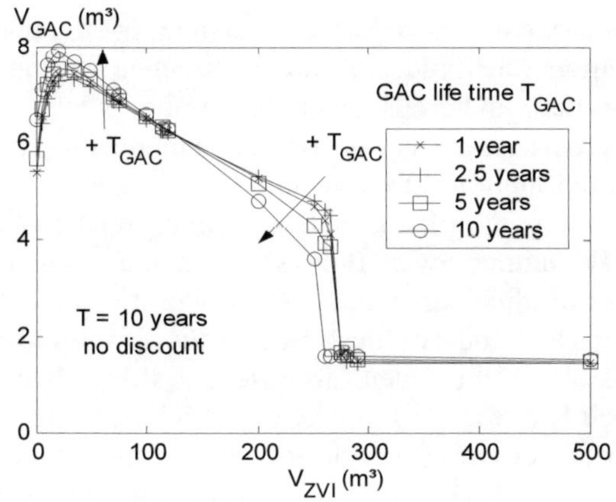

Fig. 5 Summarized total volume of GAC, V_{GAC}, for operation time of 10 years depending on the number of fills (i.e. life time of GAC reactor). Results given for the combined reactor.

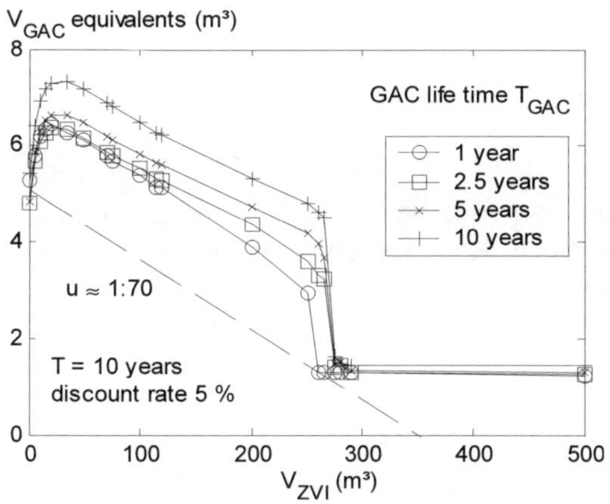

Fig. 6 Summarized total GAC volume equivalents for an operation time of 10 years if discounting is considered (see Fig. 5 for results without discounting). Dashed line represents cost isoline for a given ratio of unit ZVI costs and unit GAC costs of 1:70.

GAC per m^3) can be derived which distinguishes between two alternative reactor configurations. If *r* is below *u*, the use of a single GAC unit is favourable. Higher values yield a combined system as the optimum, which should be designed such that the least sorbing compound (*cis*-DCE) is not critical at the GAC reactor outlet. For the selected example, *u* scores 1:70 which virtually prefers the use of solely GAC.

CONCLUSIONS

The results for the $V_{ZVI} - V_{GAC}$ trade-offs for the selected example case illustrate the interference of ZVI- and GAC-specific processes. Chlorinated ethenes can be degraded by ZVI or sorbed in the subsequent GAC reactor. Enquiring which material is ideal to be used for each specific contaminant, apparently, the answer is governed by the given (relative) values of degradation rates and sorption parameters. If the treatability of contaminants is similar for both materials, such as in the presented study, some secondary reactor-specific processes appear to be relevant design criterions. In the selected example both treatment options are less suited for *cis*-DCE than for the other ethenes, denoting that the occurrence of *cis*-DCE as a byproduct in the ZVI reactor determines required material volumes even if it is not in the influent groundwater. Since the sorption capacity of this contaminant is lower than for the competing compounds, accumulated *cis*-DCE may be desorbed in the GAC and exhibit critical breakthrough concentrations above the given target levels although its concentration was uncritical at the interface between ZVI and GAC. Obviously, the combination of ZVI and GAC would be more effective if opposite treatability trends exist, e.g. if degradation rate constants of the compounds decline and sorptivity rises with increasing chlorination.

Since the modelling results considerably depend on the material-specific conditions, the derived economically optimal system is only valid for the presented individual case. The demonstrated concept based on unit prices, however, is a straightforward and generally applicable method to identify low cost options. In practice, further aspects such as expenditures for reactor construction and material replacement need to be considered.

Acknowledgements Funding for this study has been provided by the German Department of Education, Science, Research and Technology (BMBF), Contracts no. 02WT0019 and no. 02WR0195.

REFERENCES

Arnold, W. A., Roberts A. L. (2000) Pathways and kinetics of chlorinated ethylene and chlorinated acetylene reaction with Fe(0) particles. *Environ. Sci. Technol.* **34**(9), 1794–1805.

Bayer, P. (2003) Modelling, economic assessment and optimisation of in-situ groundwater remediation systems. PhD Thesis, Center for Applied Geosciences, University of Tuebingen., Germany.

Bell, L. S. J., Devlin, J. F., Gillham, R. W. & Binning, P. J. (2003) A sequential zero valent iron and aerobic biodegradation treatment system for the degradation of nitroaromatics. *J. Contam. Hydrol.* **66**, 201–217.

Crittenden, J. C., Hutzler, N., J., Geyer, D., Oravitz, J. L. & Friedman, G (1986) Transport of organic compounds with saturated groundwater FLOW: model development and parameter sensitivity. *Water Resour. Res.* **22**(3), 271–284.

Fiorenza, S., Oubre, C. L., Ward, C. H. (2000) *Sequenced Reactive Barriers for Groundwater Remediation.* Lewis Publishers, Boca Raton, Florida, USA.

Köber, R., Schäfer, D., Ebert, M. & Dahmke A. (2001) Coupled *in-situ* reactors using Fe^0 and activated carbon for the remediation of complex contaminant mixtures in groundwater. In: *Groundwater Quality: Natural and Enhanced Restoration of Groundwater Pollution* (ed. by S. F. Thornton & S. E. Oswald) (Proc. Third Int. Conf., Sheffield, UK), 435–439. IAHS Publ. 275. IAHS Press, Wallingford, UK.

Kouznetsova, I., Bayer, P., Birk, S., Liedl, R. & Finkel, M. (2003) Numerical modelling of the sequenced use of zero valent iron and activated carbon for *in-situ* groundwater remediation. In: *FZK/TNO Conference on Contaminated Soil* (Proc. Eighth Int. Conf., Gent, Belgium), 932–941.

Tiehm, A., Schulze, S., Böckle, K., Müller, A., Lorbeer, H. & Werner, P. (2000) Elimination of chloroorganics in a reactive wall system by biodegradation on activated carbon. In: *FZK /TNO Conference on Contaminated Soil* (Proc. Seventh Int. Conf., Leipzig, Germany), 924–931.

Wüst, W. F., Köber, R., Schlicker, O. & Dahmke, A. (1999) Combined zero- and first-order kinetic model of the degradation of TCE and cis-DCE with commercial iron. *Environ. Sci. Technol.* **33**(23), 4304–4309.

Feasibility study of a polymer based diffusion delivery system for ethanol to promote reducing conditions and enhanced bioremediation of contaminated groundwater

MICHELLE E. GRASSI[1, 2], BRADLEY M. PATTERSON[1], BLAIR S. ROBERTSON[1], GREG B. DAVIS[1] & ALLAN J. MCKINLEY[2]

1 *CSIRO Land and Water, Private Bag No. 5, Wembley, Western Australia 6913, Australia*
Michelle.Grassi@csiro.au

2 *Chemistry M313, School of Biomedical & Chemical Sciences, The University of Western Australia, 35 Stirling Highway, Crawley, Western Australia 6009, Australia*

Abstract The use of hollow-fibre silicone (poly(dimethylsiloxane)) tubing coiled and shaped as mats has been evaluated for its potential to act as an *in situ* permeable reactive barrier (PRB) for the controlled and targeted delivery of ethanol, to promote reducing conditions and enhanced bioremediation of a range of contaminants in groundwater. Aqueous ethanol solutions were recirculated through the inner volume of the silicone polymer tubing to allow permeation and delivery of ethanol by diffusion through the tubing wall to a target contamination zone. Laboratory and field experiments showed similar ethanol mass flux delivery which was consistent with model predictions. This study also shows promising laboratory and field simulations of this innovative amendment delivery system to predictably deliver ethanol as a carbon source for the promotion of *in situ* reducing reactive zones for the bioremediation of nitrate and metal contaminated groundwater.

Key words diffusion delivery; ethanol; groundwater; mats; permeable reactive barrier; polymer tubing; remediation

INTRODUCTION

In situ permeable reactive barriers (PRBs) can be created from hollow-fibre polymer tubing, which has the ability to deliver aqueous soluble amendments to the subsurface via diffusion. Polymer mats made from silicone tubing coiled into frames (Fig. 1 (insert)), have been previously used to deliver oxygen to simulate *in situ* biodegradation of atrazine in groundwater in large-scale soil column experiments (Patterson *et al.*, 2002a). Further studies have shown that silicone tubing polymer mats may promote *in situ* bioremediation of ammonium and nitrate contamination, via sequential delivery of oxygen (by a first mat in the groundwater flow path) for nitrification, and then ethanol further downgradient (by a second mat) for denitrification (Patterson *et al.*, 2002b).

As with all PRBs, this polymer tubing amendment delivery system relies on the natural hydraulic gradient of the aquifer to induce the groundwater flow past the tubing (mat) barrier. This has the potential for cost effective remediation compared to long term costs involved in groundwater pumping and above ground treatment (Davis & Patterson, 2003). The polymer tubing mats allow for continuous delivery of amendments to target contaminants in the ground over sustained periods of time

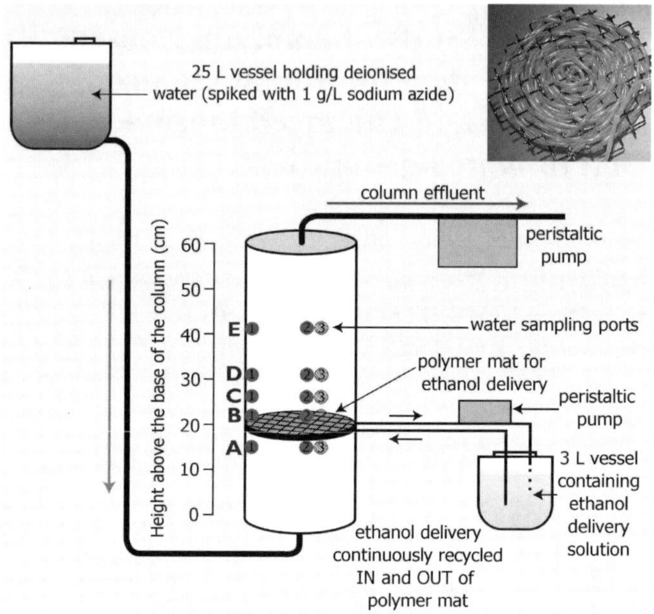

Fig. 1 Schematic of the flow through column used (insert: column scale polymer mat).

without mat degradation, eliminating the potential cost of re-installing or regenerating other types of PRBs such as ground rubber (Kershaw *et al.*, 1997), sawdust (Schipper & Vojvodić-Vuković, 2000) or zero valent iron (Gillham & O'Hannesin, 1994). With careful placement and operation of the polymer mat delivery system, amendment delivery has the potential to be controlled to meet the demand of contaminants in groundwater. However, for controlled delivery the diffusive mass transport properties of amendments through the specific polymer tubing used must be understood. This paper quantifies these processes for ethanol delivery via a silicone tubing polymer mat delivery system to promote enhanced bioremediation in groundwater.

METHODS

Model

A model has been developed from a Fickian diffusion equation (Crank, 1975) to describe the mass of ethanol delivered from silicone tubing in a polymer mat to surrounding groundwater flowing past the mat, and is described by:

$$C_1 = \frac{C_2 f_v}{z\,2\pi a}\left[\frac{z\,2\pi a}{f_v} + \frac{a\ln\left(\frac{a}{b}\right)}{D_p K_{pw}}\right] \tag{1}$$

where C_1 is the average ethanol concentration of the solution recirculating through the inside of the silicone tubing in the mat (g l^{-1}), C_2 is the ethanol concentration on the outside surface of the tubing (g l^{-1}), f_v is the volumetric flow rate of groundwater moving past the mat (cm^3 s^{-1}), D_p is the diffusion coefficient of ethanol in silicone (cm^2 s^{-1}), K_{pw} is polymer–water partitioning coefficient for ethanol in silicone (dimensionless), z is the length of silicone tubing in the mat (cm), a is the outer radius and b is the inner radius of the silicone tubing (cm). The model describes the steady-state diffusive mass flux of ethanol through the silicone tubing walls when an ethanol solution of concentration, C_1, is circulated through the inner volume of the tubing. The model accounts for the transport of ethanol away from the silicone tubing due to the groundwater volumetric flow past the polymer mat. The model assumes that the velocity of ethanol solution circulation through the inner volume of the tubing is fast enough such that only negligible concentration decreases occur along the length of the tubing. The specific details of the model development can be found in Grassi *et al.* (2004). The K_{pw} for ethanol in the silicone tubing has been previously determined as 0.037 from mass analysis of the ethanol sorbed by the polymer tubing at steady state. The D_p of ethanol in silicone tubing has also been determined from earlier analysis of transient ethanol desorption profiles from the polymer tubing into a gaseous phase, and ranged from 1.64 to 3.78×10^{-6} cm^2 s^{-1}.

Model calibration via sterile water column experiment

To test and verify the model, an initial column experiment was carried out under sterile conditions (NaN_3 1.0 g l^{-1}) in a water column (PVC pipe, 60.0 cm long, 153 mm i.d., 160 mm o.d.) (Fig. 1). The water column was operated at room temperature (~25°C), with sterile deionized water flowing in upward mode through the column. The flow was controlled by a peristaltic pump on the effluent line to a constant measurable flow (ranging between 0.30 to 0.93 ml min^{-1}). A single polymer mat (Fig. 1(insert)), consisting of 5 m length of silicone tubing (2 mm i.d., 3 mm o.d.) woven through a plastic mesh support frame (13.5 cm diameter), was installed 20 cm from the base of the water column. The polymer tubing contained a coiled stainless steel wire to prevent twisting and collapse. An aqueous ethanol solution (varying from 8 to 242 g l^{-1} ethanol) was continuously recycled through the inner volume of the tubing in the polymer mat at a rate of ~2 ml min^{-1}. Steady-state ethanol concentrations external to the tubing, C_2, were determined for a range of conditions by modifying the column water flow rates, f_v, and the ethanol concentration within the silicone polymer tubing, C_1.

Model confirmation by comparison with biological soil columns and field demonstration

The ethanol mass delivery was assessed in bioremediation/bioprecipitation large-scale soil column experiments and in a field-scale demonstration.

Two sets of large scale column experiments were conducted where ethanol was delivered using the polymer mats to promote: (a) simultaneous sulfate reduction and metal sulfide precipitation for remediation of metal/sulfate contaminated groundwater; and (b) reductive dechlorination of trichloroethene (TCE) contaminated groundwater.

The columns were constructed using a similar method to that described for the sterile water column, except that the columns were longer (2 m) and were packed with soil from known contaminated sites in Perth, Western Australia. The groundwater pumped through the columns (constant volumetric flow rate ~0.25 ml min^{-1}) was also collected from those sites. An additional mass of dissolved metals and TCE were spiked into the groundwater. The soil and groundwater in the columns were used without bioaugmentation. The same silicone tubing mat design and method of ethanol delivery was used as that for the sterile water column experiment.

Ethanol delivery by larger scale silicone tubing mats was also tested at a contaminated field site near Perth, Western Australia (Patterson *et al.*, 2004). The field demonstration consisted of a large ethanol delivery polymer mat (288 m of silicone tubing (5.5 mm o.d., 4.0 mm i.d.) woven into a plastic mesh frame (0.65 m wide, 4.5 m deep), Fig. 2(a)), installed at the centre of a gravel packed flow-through box, which was between two 10-m impermeable wings (Fig. 2(b) and (c)).

RESULTS AND DISCUSSION

Sterile water column—model calibration

The mass flux, X (g cm l^{-1} s^{-1}), given by equation (2), describes the mass of ethanol delivered per unit time per unit surface area of tubing in the mat, and can be related to C_1 as described by equation (1). Mass flux:

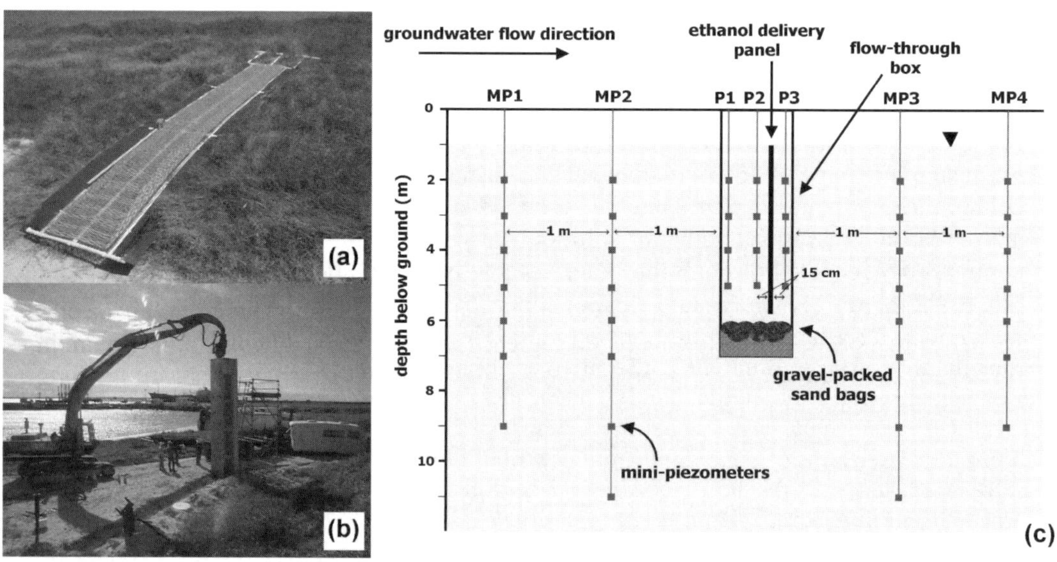

Fig. 2 (a) Field-scale ethanol delivery polymer mat panel. (b) Field site showing the layout of the treatment wall prior to installation, with flow-through box lifted and about to be installed at the centre. (c) Cross-sectional diagram of the treatment wall, showing locations of ethanol delivery panel and monitoring panels (P1, P2, and P3) within the flow-through box. Monitoring panels (MP1, MP2, MP3, and MP4) upgradient and downgradient of the flow-through box are also shown. Not to scale in the horizontal direction.

$$X = \frac{C_2 f_v}{z\, 2\pi\, a} \tag{2}$$

For each steady-state experimental data set collected from the sterile water column, X was plotted against its corresponding C_1 value (Fig. 3(a)). The model was also plotted in Fig. 3(a) for C_1 as a function of X using equation (1), where the parameter values used in the model were equivalent to the experimental conditions used in the sterile water column experiment (listed in Table 1). The model was plotted twice using the two different D_p values for the limits of the range of the diffusion coefficient (1.64 to 3.78×10^{-6} cm^2 s^{-1}).

The experimental data from the sterile water column was not well represented by the model for the assumed values of the diffusion coefficient. The diffusion coefficients were previously determined from the depletion of ethanol from the outer wall of a polymer tube using fast flowing air on the outer polymer wall. Water flow rates used in the sterile water column were lower and more typical of subsurface groundwater conditions. This lower water flow rate past the outer polymer wall may have resulted in mass transfer limitations due to poor mixing (as observed by Reynolds

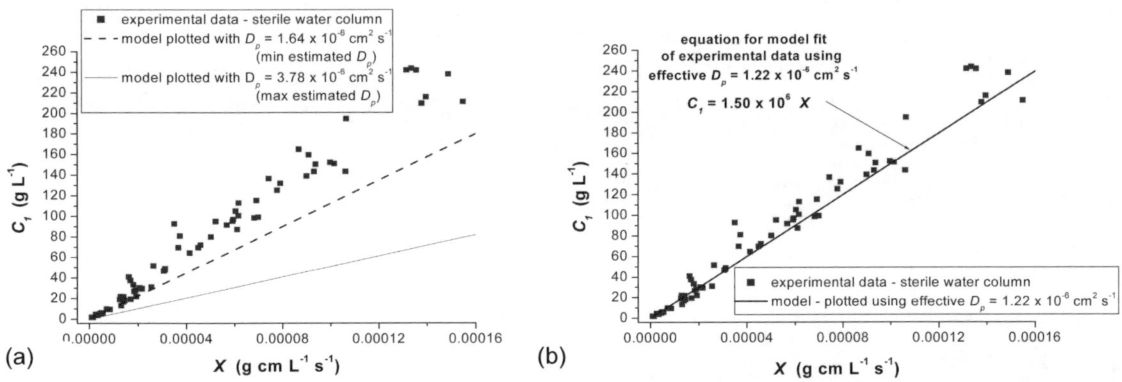

Fig. 3 (a) Graph of the experimental data from the sterile water column experiment and model (plotted twice for the range of D_p values), showing the mass flux of ethanol delivered by the tubing in the polymer mat at steady-state ethanol concentration, X, as a function of the concentration of ethanol inside the silicone tubing in the mat, C_1. (b) Graph of the refined model (using effective D_p value), showing X as a function of C_1.

Table 1 Parameters used to plot the model for the diffusive mass flux of ethanol through the walls of the silicone polymer tubing, described by equation (1).

Parameter	Value
z	500.0 (± 0.5) cm
b	0.103 (± 0.002) cm
a	0.157 (± 0.002) cm
Kpw	0.037 (± 0.006) α
fv	1.11 (± 0.08) × 10^{-2} cm^3 s^{-1} β
Dp	1.64 × 10^{-6} cm^2 s^{-1} to 3.78 × 10-6 cm^2 s^{-1} χ

α This value was determined from previous experiments. β This value is the average volumetric water flow rate through the sterile water column out of all the data collected for different experimental conditions. χ This range of values is the limits of the range previously determined for the diffusion coefficient.

et al. (1990)). The effective diffusion coefficient measured under these circumstances would be underestimated. Another possible reason for the mismatch between the experimental data and the model may be that the volumetric flow rate of water past the polymer mat was underestimated due to a reduced porosity on the pathway of flow through the mat compared to the column. However, a model sensitivity analysis indicated that the ethanol mass flux rates were relatively insensitive to the water volumetric flow rate. From fitting the experimental data to the model, an effective D_p of 1.22×10^{-6} cm^2 s^{-1} was obtained (Fig. 3(b)), a ~26% reduction in D_p from the previously estimated minimum. Despite the sensitivity of the slope of fit to the value of D_p, this percentage change would commonly be within the experimental error when estimating values of D_p. This effective D_p was used for prediction of ethanol mass delivery in subsequent column/field experiments.

Biological large-scale soil columns and field comparison

Analysis of the water collected along the length of the soil columns during ethanol delivery allowed pseudo steady-state concentrations (stable concentrations with time) of different species along the column lengths to be determined (Figs 4 and 5, for the metal bioprecipitation and TCE reductive dechlorination columns, respectively). These large-scale column experiments verified that this method of ethanol delivery has the ability to promote *in situ* denitrifying, sulfate-reducing, and metal bioprecipitation

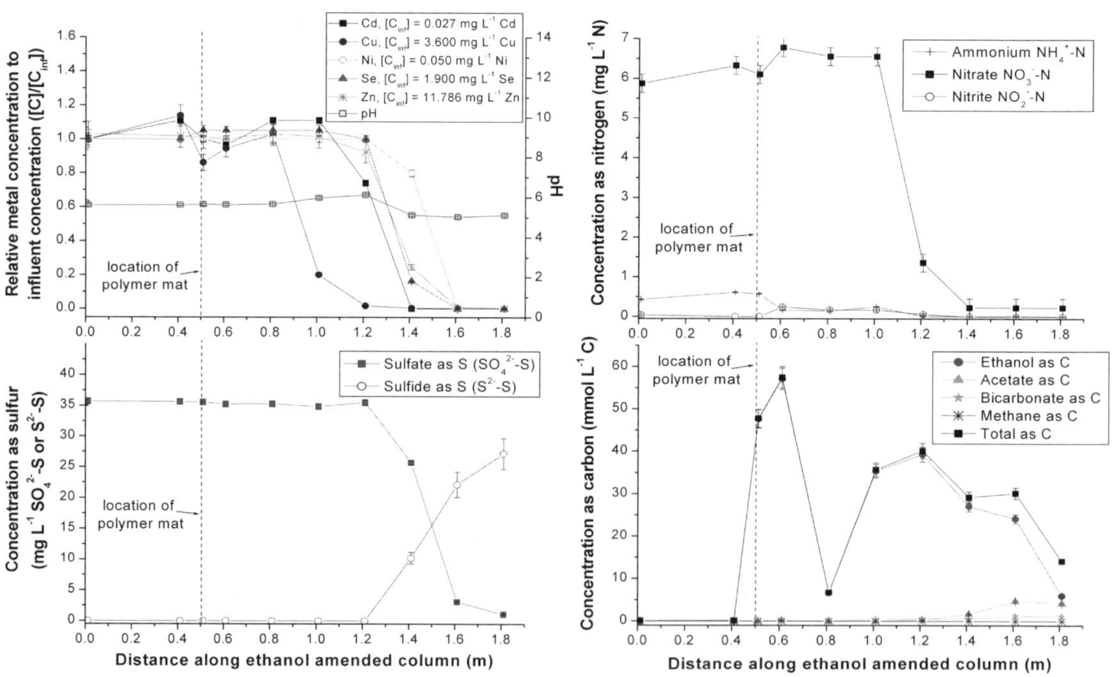

Fig. 4 Concentration data along the length of the metal bioprecipitation column at pseudo steady-state. Evidence of denitrification, sulfate reduction and metal sulfide precipitation.

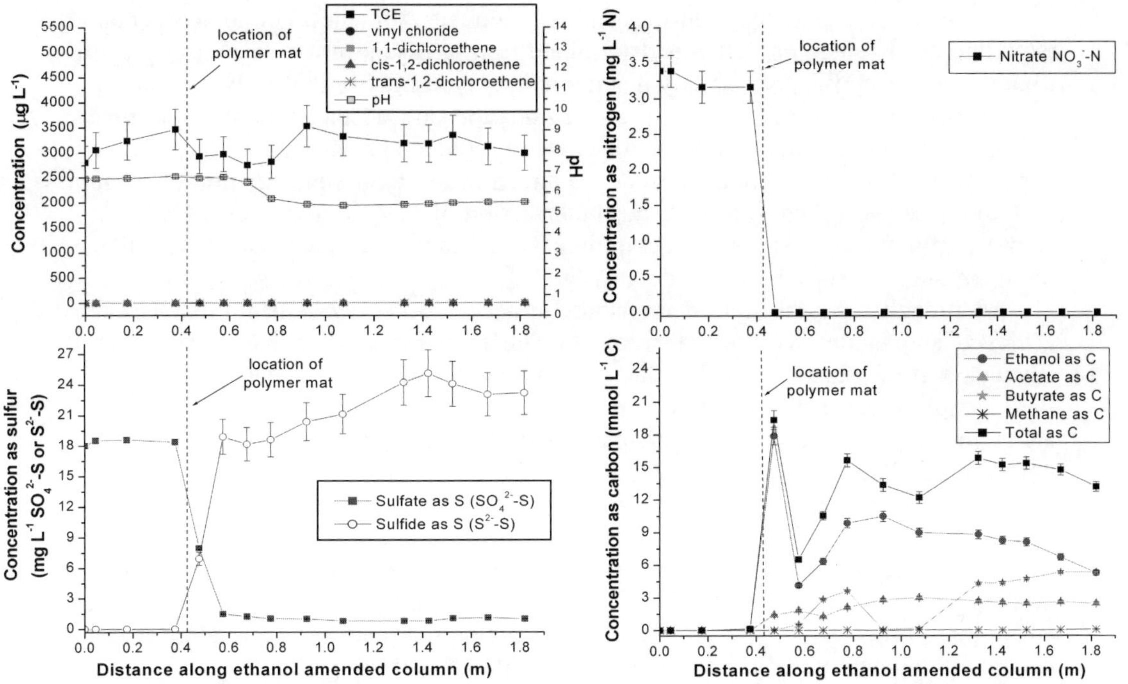

Fig. 5 Concentration data along the length of the TCE reductive dechlorination column at pseudo steady-state. Evidence of denitrification and sulfate reduction.

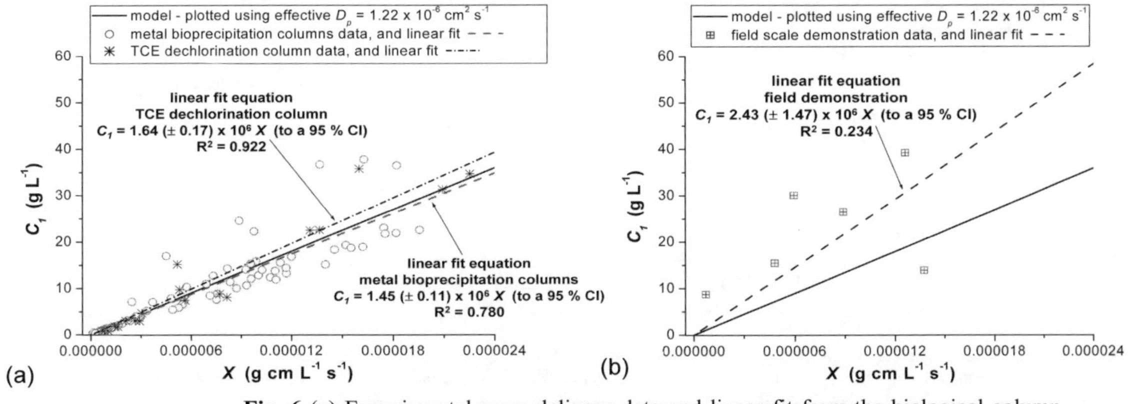

Fig. 6 (a) Experimental mass delivery data and linear fit from the biological column experiments, showing X as a function of C_1. (b) Mass delivery data from the pilot field demonstration of the silicone mat ethanol delivery system, showing X as a function of C_1. Included in both (a) and (b) is the data from the sterile water column experiment and the resulting model developed (using effective D_p value).

zones, downgradient of a polymer mat PRB. However, while denitrification and sulfate reduction was observed, no significant TCE removal in the reductive dechlorination column was been observed over a 12-month period and it was assumed that dechlorinating microorganisms were not present/active.

For every analysis day during ethanol delivery for these biological column experiments, X (equation (2)) was determined from analysis of the ethanol concentration, C_2, at the sample port directly downgradient (1–5 cm) from the polymer mat. Due to evidence of some microbial ethanol degradation and acetate formation occurring between the polymer mat and the first sample port at some stages during these column experiments, X values included acetate produced in the total concentration of ethanol measured, C_2 (assuming 1:1 molar ratio conversion of ethanol to acetate). Figure 6(a) shows X plotted against the concentrations of ethanol measured inside the silicone tubing in the mat on each sample day, C_1 (g l^{-1}).

Additionally, for the field demonstration, successful *in situ* ethanol delivery was achieved and evidence of nitrate removal due to denitrification has been reported (Patterson *et al.*, 2004). The X values (equation (2)) were also determined for each field sampling day from analysis of the average ethanol concentration (plus acetate) along the depth of the monitoring panel directly downgradient (15 cm) from the polymer mat. Figure 6(b) shows X plotted against C_1 for each particular sample day.

A linear relationship was fitted through each ethanol mass delivery data set for the column experiments and field demonstration shown in Fig. 6(a) and (b). The gradients of each of the linear relationships created were not statistically different from the model line gradient (1.50×10^6 s cm^{-1}) based on a Student's *t*-test, suggesting that the ethanol mass flux deliveries from the silicone tubing to an aqueous phase under more dynamic biologically active conditions were within the range predicted by the model. The similarity of the model estimates to the ethanol mass flux from the silicone tubing (as a function of C_2) observed in the biological large-scale column experiments and in the field scale demonstration suggest that control and prediction of ethanol delivery using silicone polymer mats is feasible and shows consistency at a range of laboratory and field scales.

Acknowledgments The authors wish to thank Nathan Innes from CSIRO Land and Water for providing technical assistance for this work.

REFERENCES

Crank, J. (1975) *The Mathematics of Diffusion*, 2nd edn. Oxford University Press, New York, USA.

Davis, G. B. & Patterson, B. M. (2003) Developments in permeable reactive barrier technology. In: *Bioremediation: A Critical Review* (ed. by I. Singleton, M. G. Milner & Head, I. M.), 205–226. Horizon Scientific Press, Wymondham, UK.

Gillham, R. W. & O'Hannesin, S. F. (1994) Enhanced degradation of halogenated aliphatic by zero valent iron. *Ground Water* **32**(6), 958-967.

Grassi, M. E., Patterson, B. M., Robertson, B., Davis, G. B. & McKinley, A. J. (2004) Characterisation of the mass delivery from an ethanol diffusion delivery system for the bioremediation of contaminated groundwater (in preparation).

Kershaw, D. S., Kulik, B. C. & Pamukcu, S. (1997) Ground rubber: sorption media for ground water containing benzene and o-xylene. *J. Geotech. and Geoenvir. Engng* **123**(4), 324–333.

Patterson, B. M., Franzmann, P. D., Davis, G. B., Elbers, J. & Zappia, L. R. (2002a) Using polymer mats to biodegrade atrazine in groundwater: laboratory column experiments. *J. Contam. Hydrol.* **54**, 195–213.

Patterson, B. M., Grassi, M. E., Davis G. B., Robertson, B. & McKinley, A. J. (2002b) Use of polymer mats for sequential reactive barrier remediation of ammonium-contaminated groundwater: laboratory column evaluation. *Environ. Sci. Technol.* **36**, 3439–3445.

Patterson, B. M., Grassi, M. E., Robertson, B., Davis, G. B. & McKinley, A. J. (2004) The use of polymer mats in series for sequential reactive barrier remediation of ammonium-contaminated groundwater: field evaluation. *Environ. Sci. Technol.* (submitted)

Reynolds, G. W., Hoff, J. T. & Gillham, R. W. (1990) Sampling bias caused by materials used to monitor halocarbons in groundwater. *Environ. Sci. Technol.* **24**, 135-142.

Schipper, L. A. & Vojvodić-Vuković, M. (2000) Nitrate removal from groundwater and denitrification rates in porous treatment wall amended with sawdust. *Ecol. Engng* **14**, 269–278.

Permeable Reactive Barriers (Proceedings of the International Symposium held at Belfast, Northern Ireland, March 2004). IAHS Publ. 298, 2005.

Enhanced microbial degradation of chloroethenes and chloroethanes in a bioscreen

NANNE HOEKSTRA[1], ALETTE LANGENHOFF[1], THEO VERHEIJ[2], JOB DIJKHUIS[3] & HANS SLENDERS[1]

1 *TNO Environment, Energy and Process Innovation, Department of Environmental Biotechnology, PO Box 342, 7300 AH Apeldoorn, The Netherlands*
n.k.hoekstra@mep.tno.nl

2 *DAF Trucks, The Netherlands*

3 *BAM-NBM, The Netherlands*

Abstract A pilot remediation with a biological treatment zone is performed at a site contaminated with chloroethenes and chloroethanes. This paper describes two different water circulation systems that were installed at the site for the infiltration of various compounds to enhance the degradation of the contaminants. Our experiences in the operation and performance of both bioscreen systems are given. Elaborate monitoring—of contaminants inclusive degradation products, hydrogen, sulphate and fatty acids—was essential to adjust the operation of the two bioscreens and the electron donor infiltration regime. The results show that both chloroethenes and chloroethanes are degraded in the two bioscreens. In the best working bioscreen, concentrations of *cis*-DCE dropped from 15 mg l^{-1} to 10 μg l^{-1}. VC concentrations initially increased from 1500 to 3300 μg l^{-1}, followed by a decrease to less then 1 μg l^{-1}. Concentrations of 1,1-DCA and 1,2-DCA dropped from 1 mg l^{-1} to 100 μg l^{-1} and 10 μg l^{-1}, respectively.

Key words bioscreen; chlorinated solvents; groundwater remediation

INTRODUCTION

Under anaerobic conditions micro-organisms are able to convert chloroethenes and chloroethanes to less hazardous compounds via reductive dechlorination. During the degradation process, the chlorine atoms are replaced by hydrogen in several steps (Middeldorp *et al.*, 1999). Reductive dechlorination is often hindered by the lack of sufficient electron donors. Addition of an electron donor may be enough to stimulate the degradation process. Different electron donors are possible as long as they are able to generate hydrogen in the soil, since this is used as the ultimate electron donor (Ter Meer *et al.*, 1999).

A contaminated site in The Netherlands has been used to study the *in situ* degradation of chloroethenes and chloroethanes. In previous studies, the dechlorination of a mixture of chlorinated ethenes and ethanes was demonstrated in batch experiments (Slenders *et al.*, 2002). The theoretical effectiveness of a bioscreen and natural attenuation on the development of the plume in the downstream aquifer has been determined with a solute transport model (RT3D). Without a bioscreen, a stationary plume will arise in 60 years and stay over hundreds of years. With a bioscreen, it is predicted that the plume will fade away in 60–80 years (Slenders *et al.*, 2002).

This paper describes the results of a current study, in which two different types of bioscreens for the infiltration of electron donors were designed, installed at the site and operated for three years.

MATERIALS AND METHODS

Site description

The contaminated site is an industrial facility in Eindhoven in the south of The Netherlands where chlorinated solvents have been used to degrease metals since the late 1940s. This has resulted in a serious soil contamination with trichloroethene (TCE) and trichloroethane (TCA). The source of the contaminants is located underneath a production facility and the plume is moving away from the contaminated source area in the direction of a nearby canal and an urban area.

The top 30 m of the soil consists of a complex sequence of sand, peat, silt and clay. Contaminated groundwater infiltrates in a cork-screw shaped flow path through the top layer. It reaches the underlying aquifer at a depth of 30 m and flows underneath the canal. In the aquifer a contaminant plume is formed with a length of approximately 700 m. A geochemical and redox characterization with hydrogen measurements has indicated that natural dechlorination stagnates at a depth of 10–20 m below ground surface (b.g.s.) (Slenders *et al.*, 2002). In this soil layer the degradation product *cis*-dichloroethene (*cis*-DCE) is the main pollutant, with concentrations of 10–70 mg l^{-1}. Other important pollutants are vinyl chloride (VC) with concentrations of 2–5 mg l^{-1} and dichloroethanes (DCA) with concentrations of 1–1.5 mg l^{-1}. The bioscreens were created at this part of the contaminant plume.

Bioscreen design

Two different configurations of infiltration and extraction screens were installed to test the optimal lay-out for a durable and cost-effective injection and mixing of electron donors. Both bioscreen systems contain one extraction and two infiltration filters, but vary in the positioning of the filters, perpendicular and parallel to the direction of the groundwater flow (Fig. 1), and the screened intervals of the filters (Fig. 2).

Bioscreen I The extraction filter is situated between the infiltration filters and the treated water is recirculated regularly. Each filter has a continuous screen installed over three permeable soil layers between 10 and 20 m b.g.s. and is blinded at the clay and peat layers. Groundwater from the permeable layers is extracted, combined, electron donor is added above ground and then the resulting mix is re-infiltrated. As a result the concentrations of the chlorinated compounds, electron donor and geochemical parameters are levelled out in this bioscreen.

Bioscreen II The extraction filter is located upstream from the infiltration filters, at an adequate distance, to ensure that polluted water is infiltrated which has not been treated before. Groundwater from the permeable layer between 16 and 20 m b.g.s. is extracted, electron donor is added above ground and then re-infiltrated.

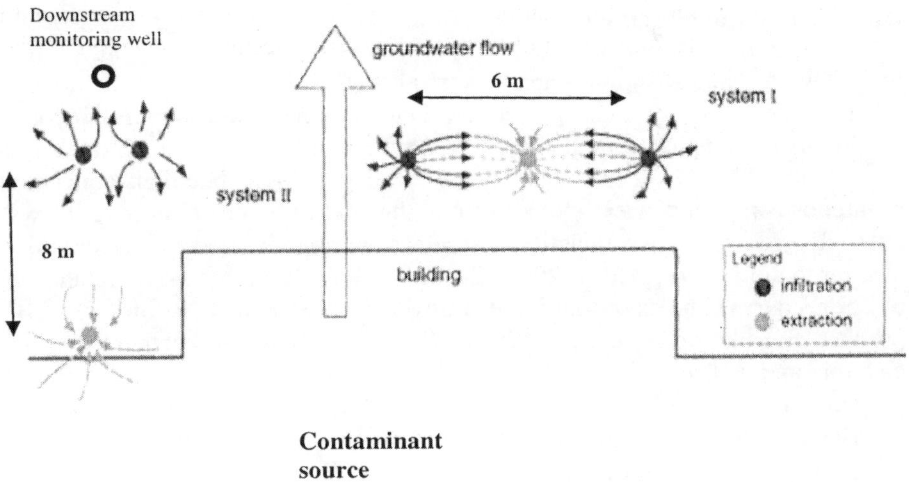

Fig. 1 Schematic presentation of the horizontal lay-out of the two bioscreen systems.

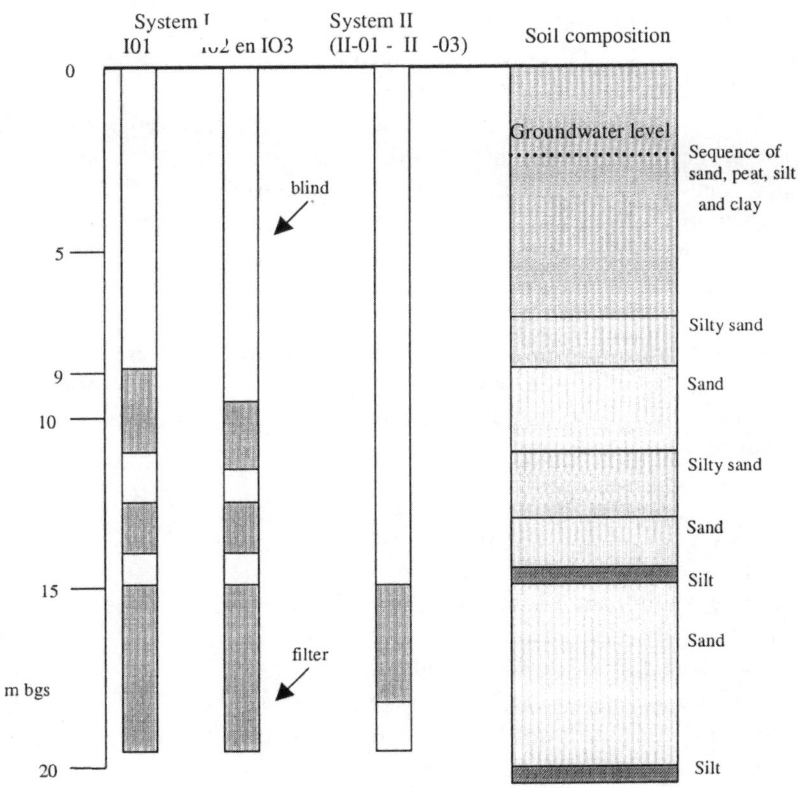

Fig. 2 Schematic presentation of the vertical lay-out of the two bioscreen systems.

Monitoring wells are located near the infiltration filters (max. 6 m distance), in order to measure the effects of operational changes quickly.

Operation of the bioscreen

In 2001 a solution of sodium-lactate was injected. A bulk donation in July 2001 of 20 times the theoretical amount of electron donor was needed to obtain the correct redox conditions (no sulphate, methanogenic). At the end of 2001 there was clear evidence (see results, Fig. 3) of enhancement of the degradation, but high concentrations of hazardous compounds were still present at the site. In order to improve the working of both bioscreens, higher quantities of electron donors were injected and different electron donors were used in 2002. Lactate was replaced by ethyl lactate (with more electron equivalents) and additional nitrogen in bioscreen I. In bioscreen II a waste product from the potato starch industry (tested positively in batches) was used as an electron and nutrient donor. This product contains high levels of trace elements, proteins, etc.

Besides this change in infiltration, the operation of bioscreen I was further adjusted. At the beginning of 2002, only a part of the desired amount of electron donor was applied in bioscreen I. An increased pressure in the injections filters indicated a reduction of the infiltration capacity due to clogging. As a result, a change in extraction and infiltration regime was implemented in bioscreen I. Groundwater was extracted from the filter on the right instead of the middle and infiltrated in the middle and left filter (Fig. 1). This change resulted in the infiltration of more water that had not been treated before and the operation of bioscreen I now resembles bioscreen II. Furthermore, the electron donor of bioscreen I was also changed to the potato starch waste product at the beginning of 2003. After three years of operation of the bioscreens approximately 4000 kg electron donor has been used.

Quantities of circulated groundwater and added electron donor as well as the frequency of the infiltration were changed several times during the test period. As a result, an optimized regime was established. For the bioscreens I and II, on a weekly basis, the amounts of extracted and re-infiltrated groundwater are now approximately 15 m^3 and 2 m^3, respectively, with electron donor additions of 30 l and 10 l. The extraction of groundwater, addition of electron donor, and re-infiltration is completed within one day.

Nowadays the optimized bioscreens are operated with remote control, reducing the time needed for inspection and maintenance.

RESULTS AND DISCUSSION

Since the start of the bioscreen operation (end of 2001), the degradation of chlorothenes has been significantly enhanced in both bioscreens. TCE disappeared and *cis*-DCE was transformed in VC and ethene. Bioscreen II performed very well (Fig. 3), but in bioscreen I the chlorinated compounds remained present for a long time. Various parameters, e.g. sulphate, redox potential, pH, nutrients and volatile fatty acids were monitored in order to explain the stagnating biodegradation. Most parameters reached the desired condition for reductive dechlorination relatively quickly, e.g. a stable redox potential of –350 mV, neutral pH, sufficient nutrients levels (results not shown). The bulk donation in 2001 was needed for the removal of sulphate (concentrations decreased from 100 mg l^{-1} to 1 mg l^{-1}), which acts as a competing electron acceptor with the chlorinated compounds for the added electron donor. In bioscreen II sulphate disappeared almost completely but in bioscreen I small residual concentrations remained. In bioscreen I, sulphide concentrations increased to more than 500 mg l^{-1} after the large injection in May 2002. High sulphide levels did not inhibit degradation in microcosm studies nor in bioscreen II. Levels of fatty acids fluctuated strongly; they increased after the high electron donor doses in 2002, but dropped in between the injections after May 2002. Hydrogen levels increased to more than 400 nM after each injection of electron donor. Initially the levels dropped quickly to less than 1 nM, resulting in insufficient conditions for dechlorination reactions. Since hydrogen concentrations are determined on-site, immediate adjustments in the electron donor doses are possible. After the increase of electron donor in 2002, the average hydrogen levels have risen to around 5 nM, sufficient for full dechlorination at the site. Low average hydrogen levels (<2 nM) coincided with stagnating degradation of chloroethenes.

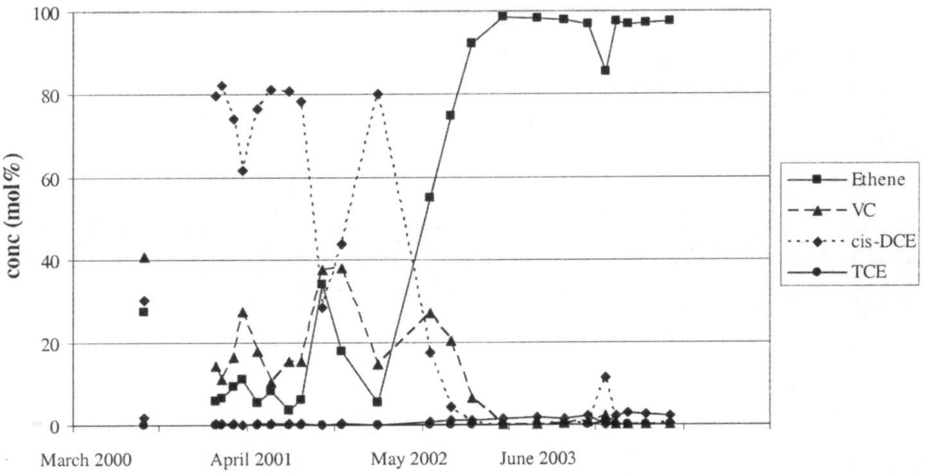

Fig. 3 Development of concentrations of chlorinated ethenes in a representative monitoring well in bioscreen II (located approximately 5 m downstream of the infiltration filters; see Fig. 1).

Increased sulphate concentrations and temporary low hydrogen levels indicated the main problem: not enough electron donor was reaching all soil layers. This was caused by clogging of the filters. It turned out that, no matter what pumping regime was chosen, bioscreen I clogged rapidly as long as it functioned as the original recirculation bioscreen. The assumption that sulphides would precipitate in the soil, where they are formed during the resting periods in between the short pumping intervals, turned out to be untrue. Microscopic investigation revealed that the re-use of amended groundwater caused the production of iron sulphide precipitates and accompanying growth of micro-organisms in the injection filters, thus decreasing the infiltration capacity of bioscreen I.

The change in configuration of the bioscreen and the added electron donor lead to a considerable increase in the performance of bioscreen I. By the end of 2003, 70% (in the less permeable layers) to 90% of the chlorinated ethenes are dechlorinated (results not shown). In bioscreen II complete dechlorination of these compounds was already established by the end of 2002 (Fig 3). In bioscreen II the *cis*-DCE concentrations decreased from 15 mg l^{-1} to 10 μg l^{-1}. VC concentrations initially increased from 1500 μg l^{-1} to 3300 μg l^{-1}, followed by a decrease to less then 1 μg l^{-1}. Currently, these typical concentrations are maintained for 1.5 year.

Some clogging of the filters by floating particles in the electron donor mixture did not affect these results. DNA-analyses showed the presence of *Dehaloccoides ethenogenes*, a bacterium that can dechlorinate TCE completely to ethene. Also *trans*-dichloroethene and 1,1-dichloroethene, which are present in relatively low concentrations, are removed effectively in the bioscreens (results not shown).

At the site, chlorinated ethanes are present in relatively high concentrations as well, e.g. 1,1-DCA and 1,2-DCA. The degradation of these compounds also started in 2003 (Fig. 4), and concentrations dropped from 1 mg l^{-1} to 100 μg l^{-1} 1,1-DCA and 10 μg l^{-1} 1,2-DCA. Recently it was proven that a *Dehalobacter* sp. can grow on 1,1,1-trichloroethane (Sun *et al.*, 2002), but DNA-analyses showed that this organism was not present at the site. At the site, DCA is transformed to chloroethane (CA). Further anaerobic degradation of CA is part of our current research.

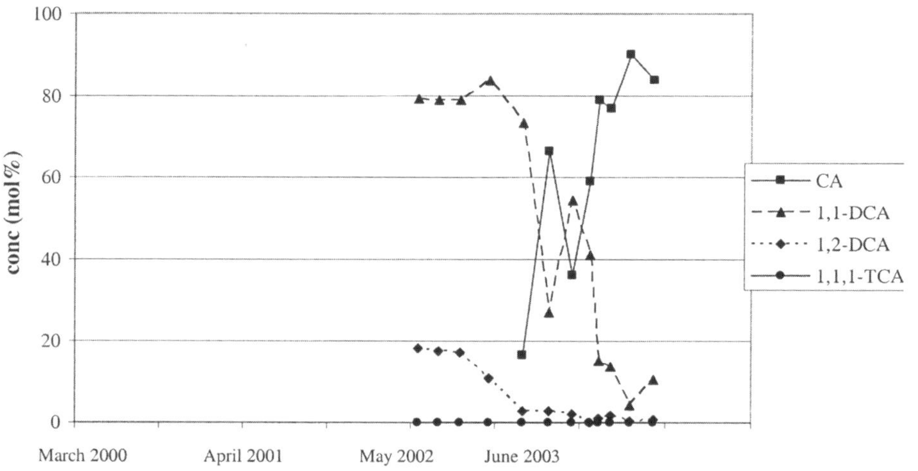

Fig. 4 Development of concentrations of chlorinated ethanes in a representative monitoring well in bioscreen II (located approximately 5 m downstream of the infiltration filters; see Fig. 1).

CONCLUSIONS

From the operation of the bioscreen and the results of the monitoring the following conclusions can be derived:

(a) A durable and efficient bioscreen for chlorinated carbons was created.
(b) The degradation of both chloroethenes and chloroethanes was enhanced successfully.
(c) A waste product of the potato starch industry is very suitable as a complex electron

donor. This molasses like product is cheap and only requires some facilities to avoid the infiltration of floating particles.

(d) The configuration of a bioscreen which involves the recirculation of already treated water causes clogging of the filters with sulphide and organic particles.

(e) Ethyl lactate can cause a severe drop in the pH and should therefore be applied with precaution. In most cases it should also be amended with nutrients and trace elements.

(f) Hydrogen measurements have proven to be very suitable for quick adjustments in the injection regime of the electron donor. Hydrogen levels should be typically between 3–6 nM.

REFERENCES

Middeldorp, P. J. M., Luijten, M. L. G. C., Van de Pas, B., Van Eekert, M. H. A., Kengen, S. W. M., Schraa, G. & Stams, A. J. M. (1999) Anaerobic microbial reductive dehalogenation of chlorinated ethenes. *Bioremediation J.* **3**, 151–170.

Slenders, J. L. A., Verhaagen, P., Hoekstra, N. K., Van Eekert, M., Dubbeldam, R. & Ter Meer, J. (2002) Intrinsic biorestoration within a large CAH-contamination, Phase 2: Additional characterization and pilot remediation. *SKB-report SN-005.*

Sun, B., Griffin, B. M., Ayal-del-Rio, H. L., Hasham, S. A. & Tiedje, J. M. (2002) Microbial dehalorespiration with 1,1,1,-trichloroethane. *Science* **298**(285), 1023–1025.

Ter Meer, J., Gerritse, J., Mauro, C. di, Harkes, M. P. & Rijnaarts, H. H. M.(1999) Hydrogen as an indicator for *in situ* redox conditions and dechlorination. *NOBIS-report 96.024.*

Permeable Reactive Barriers (Proceedings of the International Symposium held at Belfast, Northern Ireland, March 2004). IAHS Publ. 298, 2005.

Carbon disulphide sources, risks and treatment: an emerging contaminant of environmental concern?

ROBERT M. KALIN, PHILIP MYLES & KAREN McGEOUGH
Environmental Engineering Research Centre, Queen's University Belfast, Belfast BT9 5AG, UK
r.kalin@qub.ac.uk

Abstract Carbon disulphide (CS_2) is a contaminant of concern at a significant number of US EPA Superfund sites, and is found at an increasing number of contaminated land and groundwater sites. The risk assessment and remediation of CS_2 therefore requires a wider appreciation for the potential occurrences and what remediation options are available when identified as a contaminant of concern. This paper reviews the sources, risks and remediation options that are currently known for CS_2 contaminated soil and groundwater.

Key words carbon disulphide; contaminated land; groundwater remediation; permeable reactive barrier; zero valent iron

INTRODUCTION

Carbon disulphide (carbon bisulphide, dithiocarbonic anhydride), CS_2, is a dense non-aqueous phase (DNAPL) compound which has high volatility and flammability. Low concentrations of carbon disulphide occur naturally though discharge into the atmosphere from certain soils, plant extracts such as mustard oil, volcanic gases, and crude petroleum (Watts, 2000; Muezzinoglu, 2003). CS_2 is a terminal biodegradation product of CCl_4 in reducing environments in laboratory experiments (Davis *et al.*, 2003). Low and constant values of CS_2 have been shown to be released in estuaries (Moret *et al.*, 2000), and CS_2 is released during decomposition of leaf tissues of *Brassica juncea* in sandy-loam and clay-loam soils (Bending & Lincoln, 1999). Natural CS_2 concentration ranges suggest that the production and fate of CS_2 and dimethyl sulphide (DMS) in water are varied and may represent an appreciable fraction of the total natural reduced sulphur emitted to the atmosphere (Weber, 1999). Microbial respiration of the S-containing compounds CS_2, dimethyl-disulphide, dimethyl-sulphide and methanethiol are the dominant headspace components and potentially may provide a natural pesticide property to the soil (Scholler *et al.*, 1997). However, significant anthropogenic use of CS_2 has resulted in releases to the environment at concentrations that are many orders of magnitude greater than the natural background. In fact, recently this compound has been found to be one of the most common contaminants in shallow groundwater (Baehr *et al.*, 1999).

CS_2 was first prepared nearly 200 years ago by heating sulphur with charcoal. Significant commercial production of CS_2 began around 1880, primarily for agricultural and solvent applications. Production of CS_2 expanded rapidly after World War II to supply the growing requirements of the viscose rayon industry, which consumed about 0.31 ton CS_2 per ton of rayon produced. CS_2 is also used in the rubber industry to make intermediates in the manufacture of rubber vulcanization accelerators, including MTB (2-mercaptobenzothiazole), thiocarbanilide, and dimethyl dithiocarbamate salts. Thiocarbanilide is also used in dyes. Other uses of CS_2 are in the manufacture of cellophane and carbon tetrachloride, CCl_4 (Dunn *et al.*, 1989). CS_2 for the manufacture of CCl_4 increased in the 1950s and 1960s to supply key raw material for chlorofluorocarbon (CFC) refrigerants and aerosol propellants. Due to ecological and health concerns, CCl_4 consumption began to decline in the mid-1970s. The demand for CS_2 has been increased due to the phasing out of CCl_4 and CFCs to protect the Earth's ozone layer. It is also an unintentional by-product of many combustion and high temperature industrial processes where sulphur compounds are present (Gupta & Tulis, 1992; Jensen & Webb, 1995; Revah *et al.*, 1995; Hugler *et al.*, 1999). Thus, as with most chemicals used in manufacturing during the 19th and 20th centuries, significant releases to the environment have occurred (Chemical Market Reporter, 2001; NPL Site Narrative for Avtex Fibers Inc., 2001; NPL Site Narrative for Halby Chemical Co., 2001; NPL Site Narrative for Stauffer Chemical Co., 2001).

CS_2 also has a long history of use as an agricultural fumigant and it is an intentional degradation product of a group of pesticides called "dithiocarbamates" (Gallian & Hafex, 1984; Draper & Wakeham, 1993; Wijnands & van Asperen, 1999). It was used in France in 1869 as a soil fumigant to control the grape fungus, *Phylloxera*, and to control the sugar beet nematode, and became widely used as a soil and grain fumigant after the First World War (Matheron & Matejka, 1987; Young & Green, 1988). However, its volatility, high liquid and vapour toxicity, and extreme flammability have made its direct use impractical (Gallian & Hafez, 1984). The increased use of phosphine and the removal of CS_2 from the US EPA list in the 1980s led to a decline in its use. However, with the proposed limitations on the use of methyl bromide as an agricultural fumigant of choice (Kalin *et al.*, 2001), and increasing insect resistance to phosphine, there is a renewed interest in CS_2 as a fumigant as it acts relatively quickly, providing a rapid method of disinfecting the agricultural product. With on-farm storage the potential for diffuse contamination of aquifers is significant.

A lack of knowledge with respect to the risk of CS_2 in soil and groundwater has resulted in the uncertainty for drinking water standards for CS_2. In one example, the State of California (1991) suggested a level of 770 mg l^{-1} based on the US EPA Integrated Risk Information System's (IRIS) Carbon Disulfide document, revised 1 September 1990, and the California State Office of Environmental Health Hazard Assessment (OEHHA) recommended in July 2001 that the action level for CS_2 in water be set at 160 mg l^{-1}. Review of a number of CS_2 contaminated sites in the UK has found concentrations in groundwater ranging from a few mg l^{-1} to over 300 mg l^{-1}. The same data set has shown concentrations in soils from potentially contaminated sites ranged from tens of mg kg^{-1} (detection limits) to well over 5000 mg kg^{-1} in source zones. Some recent estimates for human health risk assessment of CS_2 in soil suggest potential risk at levels approaching the detection limits of measurement. Therefore, there is a need to better define the extent of CS_2 in soil and groundwater, to educate the soil and groundwater remediation practitioners about CS_2, and thus more effectively evaluate the risk of this compound in the environment.

Physical and chemical properties of carbon disulphide

Pure CS_2 is a clear, colourless liquid with a delicate ether-like odour. A faint yellow colour slowly develops upon exposure to sunlight. Low-grade commercial CS_2 may display some colour and may have a strong, foul odour because of sulphurous impurities. CS_2 is only slightly miscible with water, but it is a good solvent for a wide range of organic compounds, being completely miscible with many hydrocarbons, alcohols, and chlorinated hydrocarbons (Seidell & Linke, 1952). CS_2 will not be found as a solid in the environment; however, under extremely high pressures of about 5.5 GPa (5.4×10^4 atm) and temperatures up to 175°C, a black, solid form of CS_2 has been observed (Butcher *et al.*, 1964). Principal properties are listed in Table 1.

Table 1 Physiochemical properties of carbon disulphide.

Property	
Melting point, *K*	161.11
Latent heat of fusion, kJ kg^{-1}	57.7
Boiling point at 1 atm, °C	46.25°C
Flash point at 1 atm, °C	–30°C
Ignition temperature in air, °C	
10-s lag time	120°C
0.5-s lag time	156°C
Critical temperature, °C	273
Critical pressure, kPa	7700
Critical density, kg m^3	378
Solubility H_2O in CS_2	
at 10°C, mg l^{-1}	86
at 25°C, mg l^{-1}	142
Dielectric constant	2.641
Liquid at temperature 20°C	
Density	1.263
Specific heat	1005
Latent heat of vaporization	368
Surface tension	32.3
Thermal conductivity	0.161
Viscosity	0.367
Refractive index	1.6276
Solubility in water	2.1
Vapour pressure	39.66
Gas at temperature, 46.25°*C*	
Density	1.96
Specific heat	679
Viscosity	0.0164
Thermochemical data at 298 K	
Heat capacity, C_p°, J/(mol·K)	45.48
Entropy, S°, J/(mol·K)	237.8
Heat of formation, $H°_f$, kJ mol^{-1}	117.1
Free energy of formation, $G°_f$, kJ mol^{-1}	66.9

The low flash point temperature of –30°C at atmospheric pressure and wide flammability range of CS_2 deserve special attention during site investigation and remediation, in particular when free-phase may be encountered (Meidl, 1970; FMC Corporation, 1982; Emergency Actions Guidance, 1988). The flash point is lowered if the pressure is decreased or the oxygen content enriched. The flammability limits or explosive ranges depend on conditions of temperature, pressure, and geometry of the enclosure. Hot surfaces and electric sparks are potential ignition sources for CS_2. The ignition temperature depends on specific conditions, and values from 90 to 120°C in air have been reported (Emergency Action Guides, 1988). Oxidation products are generally sulphur dioxide and carbon dioxide:

$$CS_2 + 3O_2 \rightarrow 2SO_2 + CO_2$$

Carbonyl sulphide and carbon monoxide can also form under certain conditions. Thermodynamic calculations for reactions forming CS_2 from the elements are complicated by the existence of several known molecular species of sulphur vapour (Dunn & Rudorf, 1989). CS_2 decomposes abiotically extremely slowly at room temperature in the absence of oxidizing agents.

Risks associated with carbon disulphide

Exposure to CS_2 can occur by inhalation, coming into contact with the liquid, and by drinking water or eating foods that contain it. Ingestion of CS_2 is rare, but a 10 ml dose can prove fatal (Arena, 1976). Contact usually occurs by inhalation of the vapour, however, vapour and liquid can be absorbed through intact skin and poisoning may occur by the dermal route where repeated contact of liquid CS_2 removes protective waxes and oils (US National Technical Information Service, 1989). It has an irritative effect on eyes, lungs, and skin (NIA, 1988), and is also a potent neurotoxin (Vaccari *et al.*, 1999). Prolonged exposure to CS_2 has a negative effect on the blood vessels (US National Technical Information Service, 1989).

The immediate effects of overexposure to CS_2 vapours range from headache, dizziness, nausea, and vomiting to life-threatening convulsions, unconsciousness, and respiratory paralysis. Prolonged and repeated exposure to CS_2 vapour can affect both the central and peripheral nervous systems. Repeated exposure to relatively high concentrations has been known to cause serious neurological and psychological impairments (Mackison *et al.*, 1981). Certain workers exposed to a time-weighted average of 11 mg l^{-1} experienced headaches and dizziness, and those with an average of 186 mg l^{-1} had additional complaints of nervousness, fatigue, sleep problems and weight loss (Mackison *et al.*, 1981).

Environmental releases of carbon disulphide

Before the rise of environmental awareness, people did not understand how certain wastes might affect human health and the environment. Many wastes were released on the ground, in rivers or left out in the open. As a result, thousands of uncontrolled or abandoned hazardous waste sites were created. Some common hazardous waste sites included abandoned warehouses, manufacturing facilities, processing plants and landfills. In the United States, in response to growing concern over health and environmental risks posed by hazardous waste sites, Congress established the Comprehensive Environmental Response, Compensation, and Liability Act (CERCLA) commonly known as the Superfund Program. In 2001, of 1226 Superfund sites, 174 (14.2%) have CS_2 as a contaminant of concern (US EPA, 2001). The following is a short synopses of three US Superfund Sites affected by CS_2: Avtex Fibers, Inc., Virginia, Halby Chemical Co., Delaware, and Stauffer Chemical Co., Alabama. The authors refer the reader to extensive US EPA literature for detailed information about the remediation of these sites.

Avtex Fibers, Inc.

Avtex Fibers, Inc. had a 340-acre site on the Shenandoah River in Front Royal, Warren County, Virginia. It had been operated as a rayon-manufacturing complex by American Viscose (1940–1963), FMC Corporation (1963–1977), and Avtex Fibers, Inc. (1977–1989). Various wastes had been disposed of in unlined basins, landfills, and open piles. In 1974, FMC received a National Pollutant Discharge Elimination System permit, to discharge wastewater effluent into the South Fork of the Shenandoah River (Avtex Fibres Inc., 2001). Groundwater under the site contained excessive levels of CS_2, phenol, and arsenic, according to analysis conducted by the State. Groundwater directly across the river from the site contained excessive levels of carbon disulphide and phenol. About 1300 people, within 3 miles of the site, used wells from the contaminated aquifer, as a source of drinking water (NPL Site Narrative for Avtex Fibres, 2001).

Halby Chemical Co.

Halby Chemical Co. was a manufacturing facility in New Castle, New Castle County, Delaware. Speciality chemicals, primarily sulphur compounds, were processed in the chemical manufacturing plant from 1948 to 1980. In the early 1970s, Witco Chemical Co. acquired the operation, then sold it to Brandywine Chemical Co. in 1977. From 1948 to 1964, production wastewater and cooling water were disposed of by discharge into the unlined on-site lagoon. The lagoon was approximately two feet deep, covered 1½ acres, and drained into the Lobdell Canal before emptying into the Christina River. High levels of various organic and inorganic substances, including CS_2, are present in water and sediment samples from the lagoon, according to EPA investigation. Four municipal wells were part of an interconnected system that supplied approximately 150 000 people living two to three miles from the site (NPL Site narrative for Halby Chemical Co., 2001).

Stauffer Chemical Co.

Stauffer Chemical Co. (LeMoyne Plant) commenced operations in the early 1950s in Axis, Mobile County, Alabama. At first it manufactured CS_2. In 1964, it began to produce chlorine and caustic soda. In 1974, the plant expanded again, producing additional industrial inorganic compounds. During the 1950s and 1960s, Stauffer used an on-site landfill located east of the manufacturing facility and between the facility and the Mobile River and they reported that the landfill contained drums of wastes that may include organics, solvents, heavy metals, acids, and bases. Groundwater was the sole source of drinking water in this area. About 4000 people (employees of local industries and residents of Falco) were served by wells within three miles of the LeMoyne Plant landfill. Groundwater in the vicinity of the landfill was contaminated by heavy metals, chloroform, and CCl_4, according to the analyses by the EPA (NPL Site Narrative for Stauffer Chemical Co., 2001).

Remediation of carbon disulphide

To date there have been a number of methods developed to remediate CS_2 contaminated soil and groundwater, these include treatment with zero valent iron (ZVI), excavation and *in situ* chemical oxidation, and sorption. However, excavation of CS_2 contaminated soil can cause the carbon disulfide to ignite or explode when exposed to air. Excavation of soil to remediate CS_2 contaminated soil causes hazards associated with exposing carbon disulfide to air. Processes of remediating CS_2 contaminated soil without exposing it to air are therefore desirable.

We report here for the first time in the literature that iron, including ZVI and nano-iron, is effective as a remediation option for CS_2 (Tratnyek, 1996). There are over 600 referenced publications that show ZVI is a well known reactive media for reactions with chlorinated solvents, trace metals, explosives and radionuclides (Tratnyek, 1996). The use of ZVI been shown to be a highly effect treatment method for CS_2 at both laboratory and full-scale (Kalin & Myles, 2004; Bone *et al.*, 2005) resulting in its complete destruction and the precipitation of iron sulphide and ferri-oxyhydroxydes and ferri-carbonate. Half-lives range between 0.5 and 1.5 h, depending on the chemistry of the contaminated groundwater (Kalin & Myles, 2004) (Fig. 1).

$$Fe + H_2O + CS_2 \rightarrow FeS + FeO(OH)_x + FeCO_3$$

Proposed *in situ* methods using iron for CS_2 remediation include mechanical injection of iron additives such as granular or nano iron. A full-scale trial using soil mixing was recently reported (Kalin & Myles, 2004). *In situ* methods also include the use of reactive zones such as Permeable Reactive Barriers (Tratynek, 1996; Bone *et al.*, 2005). *Ex situ* methods include surface treatment cells into which the contaminated substance (e.g. soil) can be placed or pumped (e.g. groundwater) and treated by mechanical mixing with iron. Water may be added to ensure the pollutant is in contact with the iron (Tratynek, 1996). Leached water from treatment cells could be recirculated through an iron reactive zone for both *in situ* and *ex situ* soil treatment. Zero valent iron and nano-iron (Tratynek, 1996) appear to be a commercially viable, abiotic method for CS_2 remediation.

Transition metal cluster compounds have also been proposed as the basis for the development of a new class of homogeneous catalysts for CS_2 degradation. The ability of triosmium-hydride cluster complexes to effect the hydrogenation of small molecules

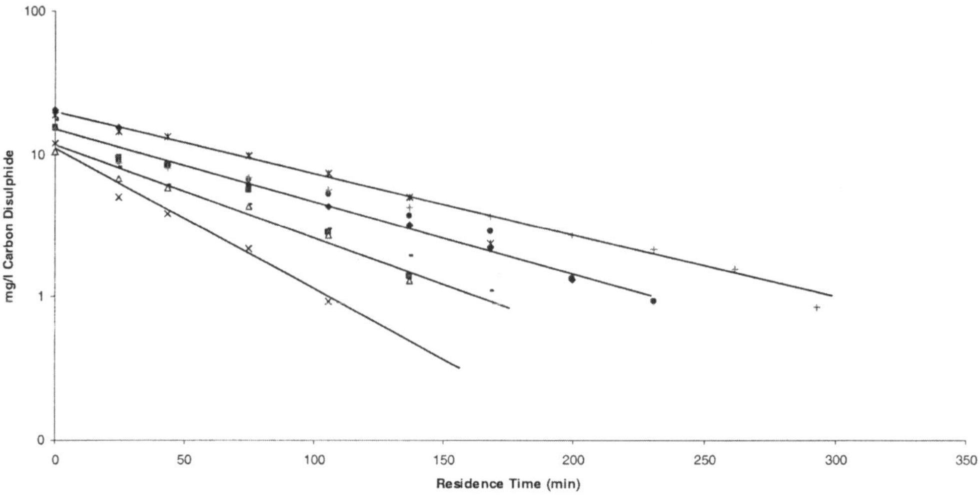

Fig. 1 Column "treatability" data from Queen's University Belfast showing degradation by zero valent iron of CS_2 from site contaminated groundwater.

had been examined (Adams *et al.*, 1979). It was found that (μ-H)2$Os_3(CO)_{10}$ reacted with CS_2 via an unusual dicluster reaction to produce the molecule (μ-S_2CH_2)[(μ-H)-$Os_3(CO)_{10}]_2$ (Adams *et al.*, 1979). It is therefore clear that cluster compounds possess remarkable abilities to hydrogenate and fragment small substrate molecules like CS_2, however there has been no effort to develop full-scale remediation applications with this chemistry.

Another possibility for the abiotic removal of CS_2 is to use modern technology to improve the efficiency of the Claus process. The Claus process was developed by Chance & Claus in 1894 and is used to remove sulphur from chemical feedstocks (Duffy & Ingram, 1971). Over the last two decades it has become apparent that recent emissions of sulphur compounds into the atmosphere have been unacceptably high. Presently the efficiency of the Claus process for the recovery of sulphur is only 90–93% of that contained in flue gases (Rhodes *et al.*, 2000). To meet the requirements of increasingly stringent pollution legislation that operates in a number of industrialized countries, it is now required that this remaining sulphur also be removed. At this stage of the processing, the residual sulphur is mainly in the form of compounds such as COS and CS_2 (Rhodes *et al.*, 2000). An alternative proposal to reduce the level of CS_2 involves hydrogenation. Using a Co-Mo-Al_2O_3 type catalyst, the sulphur-containing compounds are reduced to H_2S:

$$CS_2 + 4H_2 \rightarrow 2H_2S + CH_4$$

The composition of the catalyst used is such that the amount of cobalt is usually 3–4 times less than molybdenum, with Mo giving rise to "surface-layer" coverage of MoO_3. The cobalt is located in the upper portions of that Mo layer. Unfortunately the expense of the Co-Mo-Al_2O_3 catalyst, combined with the high temperature required for effective reduction, at present restricts commercial exploitation.

In situ chemical oxidation (US Patent; US EPA, 1998) is another potential technology for CS_2 remediation as well as the immobilization of organic compounds in hazardous wastes by the mixing of such wastes with particulate rubber and biogenic amorphous silica (US Patent). The tests conducted included waste containing CS_2, however the method does not destroy the CS_2, it is simply fixed on the additives.

Biotechnological processes, such as biofilters and bio-trickling filters, represent an alternative to the traditional abiotic methods of absorption, adsorption and thermal or catalytic oxidation for the removal of CS_2 from waste air streams (Gupta & Tulis, 1992). This is due to their wide range of applications, high removal efficiencies and reduced operating costs (Van Groenestijn & Hesselink, 1993). Biofiltration is a process that aerobically converts particular vapour phase compounds into CO_2, biomass, and water vapour (Fucich *et al.*, 1997). In this process, micro-organisms, in the form of a moistened biofilm layer, immobilized on an organic packing material, such as compost, peat, wood chips, etc., are used to catalyse beneficial chemical reactions. As a contaminated vapour stream passes through the biofilter bed, the contaminants are transferred to the biofilm and are degraded by the micro-organisms.

Desulphurization of gas streams is important from the standpoint of pollution prevention, health, safety, and corrosion. It has been shown that simultaneous removal of H_2S and CS_2 from air streams with a peat biofilter, inoculated with a *Thiobacillus* strain, is possible (Hartikainen *et al.*, 2001). Both compounds were oxidized in an acidic environment, and it was reported that both sulphurous gases at concentrations below 600 mg S m^{-3} were efficiently removed. The removal efficiencies were similar, 99%, with an empty bed retention time (EBRT) of more than 60 s. Concentrations greater than 1300–5000 mg S m^{-3} caused overloading of the filter material, resulting in high H_2SO_4 production, accumulation of elemental sulphur, and reduced efficiency. However, these results indicated that peat is suitable as a biofilter material for the removal of H_2S and CS_2 when concentrations are low.

Several biological "trickling bed filters" for gas desulphurization processes have been proposed and show potential for treating dilute sulphide-containing air streams (Jensen & Webb, 1992; Morales *et al.*, 1992; Revah *et al.*, 1995; Lobo *et al.*, 1999). There are three commercial-size plants operating with this process in Monterrey, Mexico, and one in Tonawanda, New York, USA. This process is based on a trickle-bed bioreactor, with a *Thiobacilli* consortia biofilm supported on structured packing. The reactions that are believed to occur in the system are (Smith & Kelly, 1988):

$$CS_2 + H_2O \rightarrow COS + H_2S$$

$$COS + H_20 \rightarrow CO_2 + H_2S$$

$$H_2S + 1/2O_2 \rightarrow S^O + H_2O$$

$$S^O + H_2O + 3/2O_2 \rightarrow H_2SO_4$$

The microbiological degradation of CS_2 is not ubiquitous. It was reported that only one strain of *Thiobacillus thioparus*, from eight different *Thiobacilli* studied, was able to use CS_2 (Plas *et al.*, 1993). The reaction proceeds through the formation of H_2S and carbonyl sulphide (COS) which is further transformed into H_2S and CO_2. The H_2S produced is then oxidized to elemental sulphur and subsequently to sulphuric acid. An unidentified *Thiobacillus* sp. was reported that also utilizes CS_2 (Jordan *et al.*, 1995, 1997). More recently, of four bacterial strains found near the CS_2-producing tree *Quercus lobata*, three of them *Thiobacillus*, have been isolated which could use CS_2

and a wide range of other reduced sulphur compounds as their sole energy source (Hartikainen *et al.*, 2000).

The wide range of microbial species that have a capability of CS_2 utilization suggests that for soil and groundwater remediation or risk management monitored natural attenuation (MNA) is a distinct possibility. Figure 2 shows the first field evidence of natural biodegradation of CS_2 in groundwater observed within a CS_2 plume in the UK (Kalin & Myles, 2004). MNA is a process in which indigenous or inoculated micro-organisms (e.g. fungi, bacteria, and other microbes) metabolise organic contaminants found in soil and/or groundwater, converting them to innocuous end products. It has been shown that three distinct strains (KL1, KS1, and KS2) of facultatively chemolitho-autotrophic bacteria, were able to use carbon disulphide or carbonyl sulphide as their sole energy substrates (Jordan *et al.*, 1997). These were identified as novel strains of *Paracoccus denitrificans* where autotrophic growth on carbon disulphide was confirmed by $^{14}CO_2$ fixation. Intermediates in aerobic carbon disulphide metabolism were shown to include carbonyl sulphide and hydrogen sulphide as was first demonstrated in CS_2 contaminated groundwater from field sites (Tratynek, 1996). However, anaerobic production COS and H_2S from carbon disulphide was not observed for *Paracoccus denitrificans*. It has also been shown that three types of alkaliphilic bacteria were able to utilize thiocyanate (CNS-) at pH 10. These bacteria (*Proteobacteria*, *Thioalkalivibrio*) possessed the ability to accumulate large amounts of elemental sulphur inside its cells and had the ability to oxidise CS_2 (Sorokin *et al.*, 2001). Using batch and continuous culture enrichment techniques, a novel *Eubacterium* was isolated from soil surrounding the roots of *Tagetes* plants (marigold) known producers of thiophenes. This *Eubacterium* was revealed to be a new species of *Xanthobacter*, *X. tagetidis*. *X. tagetidis* was capable of growth on C-1 compounds including the sulphur-containing C-1 compounds, carbonyl sulphide, CS_2, methanethiol and DMS (Padden, 1997). Oxygen was not essential for the successful metabolism of these compounds, thus there is some potential for natural biodegradation of CS_2 in both aerobic and anaerobic soil and groundwater conditions.

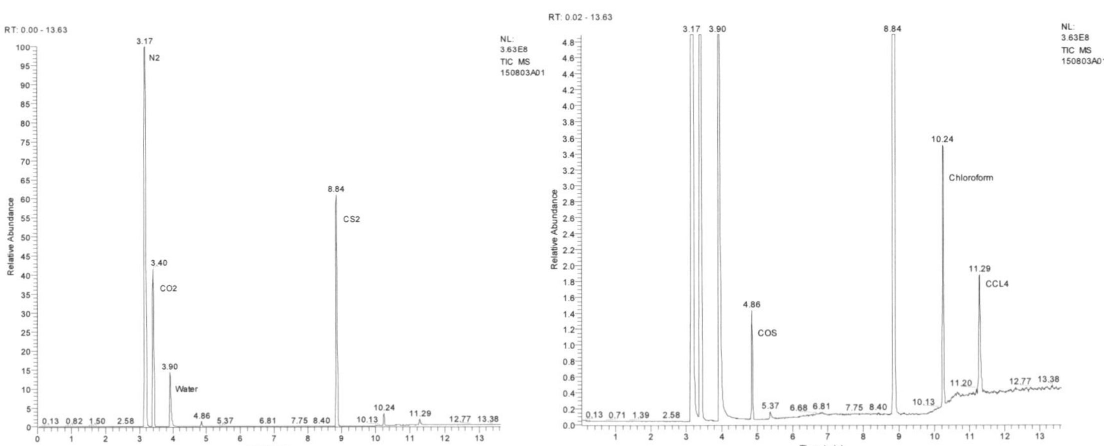

Fig. 2 GC/MS Chromatograms showing Natural Attenuation (COS) biodegradation product for CS_2 from contaminated groundwater as analysed at QUB.

CONCLUSIONS

CS_2 is an "emerging" contaminant of concern. There is a need to recognize that CS_2 may be a co-contaminant with chlorinated compounds and other solvents, behaves as a DNAPL, and has significant risk assessment requirements. However, there are a number of viable remediation technologies that exist or are emerging that provide solutions for CS_2 risk management.

Acknowledgements The authors would like to acknowledge the support of EPSRC grant GR/R03099/01, QUB SRIF investment in Environmental Engineering and Biotechnology, the Environment Agency, and industrial support from Keller Ground Engineering, CEL and ESI.

REFERENCES

Adams, R. D., Golembeski, N. M. & Selegue, J. P. (1979) The hydrogenation and degradation of carbon disulphide by organometallic cluster hydride complexes. The reaction of carbon disulphide with $Os_3(CO)_9P(CH_3)_2C_6H_5(\mu\text{-}H)_2$. *J. Am. Chem. Soc.* **101**(19), 5862–5863.

Arena, J. M. (1976) *Poisoning* (ed. by Charles C Thomas, 3rd), 84. Springfield, Illinois, USA.

Avtex Fibers, Inc. (2001) Site history: http://www.avtexfibers.com/sitehis.htm.

Bone, B. D., Harris, R. C., Smith, J. W. N., Boshoff, G., Kalin, R. M., Thurgood, R. & Morgan, P. (2005) The development and use of permeable reactive barrier technologies and potential future applications in the UK. In: *Permeable Reactive Barriers* (ed. by G. Boshoff & B. D. Bone) (Proc. Int. Symp., Belfast, Northern Ireland, March 2004), 52–59. IAHS Publ. 298. IAHS Press, Wallingford, UK (this volume).

Baehr, A. L., Stackelberg, P. E. & Baker, R. J. (1999) Evaluation of the atmosphere as a source of volatile organic compounds in shallow groundwater. *Water Resour. Res.* **35**(1), 127–136.

Bending, G. D. & Lincoln, S. D. (1999) Characterisation of volatile sulphur-containing compounds produced during decomposition of *Brassica juncea* tissues in soil. *Soil Biol. Biochem.* **31**, 695–703

Butcher, E. G., Weston, J. A. & Gebbie, H. (1964) *J. Chem. Phys.* **41**, 2554.

Chemical Market Reporter (2001) Superfund Cleanup Program, *FindArticles.com*: http://www.findarticles.com/cf_0/m0FVP/2_259/69069515/print.jhtml

Davis, D, Fennemore, G. G., Peck, C., Walker, C. R., McIlwraith, J. & Thomas, S. (2003) Degradation of carbon tetrachloride in a reducing groundwater environment: implications for natural attenuation. *Appl. Geochem.* **18**, 503–525.

Draper, W. M. & Wakeham, D. E. (1993) Rate constants for metam-sodium cleavage and photodecomposition in water. *J. Agric. Food Chem.* **41**, 1129–1133.

Duffy, J. A. & Ingram, M. D. (1971) *J. Am. Chem. Soc.* **93**, 6448.

Dunn, A. D. & Rudorf, W. D. (1989) *Carbon Disulphide in Organic Chemistry*. Ellis Horwood Ltd., Chichester, UK.

Emergency Action Guides (1988) Carbon Bisulphide. *Association of American Railroads, Bureau of Explosives, Washington, DC C2. 1(1-6).*

FMC Corporation (1982) *Carbon Disulphide: Properties, Handling and Storage*. Princeton, New Jersey, USA.

Fucich, W. J., Yang, Y., Togna, A. P. & Alibeckoff, D. (1997) Biofiltration for control of carbon disulphide and hydrogen sulphide vapours. In: *Proceedings of the Air and Waste Management Association's Annual Meeting and Exhibition.*

Gallian, J. J. & Hafez, S. L. (1984) Irrigation application of nematicides for mid-season control of sugar beet cyst nematode. *Fungicide and Nematicide Tests* **39**, 88–89.

Gupta, R. & Tulis, D. (1992) Carbon disulphide emission control options. *EPA Report 92-115.04, Research Triangle Park NC 27711: 1-22.*

Hartikainen, T., Ruuskanen, J. von Wright, K. R. A. & Martikainen, P. J. (2000) Physiology and taxonomy of *Thiobacillus* strain TJ330, which oxidizes carbon disulphide (CS_2). *J. Appl. Microbiol.* **89**, 580–586.

Hartikainen, T., Ruuskanen, J. & Martikainen, P. J. (2001) Carbon disulphide and hydrogen sulphide removal with a peat biofilter. *J. Air and Waste Manage. Assoc.* **51**(3), 387–392.

Hugler, W., Acosta, C. & Revah, S. (1999) Biological removal of carbon disulphide from waste air streams. *Environmental Progress*, **18**(3), 173–177.

Jensen, A. B. & Webb, C. (1995) Treatment of H_2S-containing gases: A review of microbiological alternatives. *Enzyme Microb. Technolo,* **17**, 2–10.

Jordan, S. L., Kraczkiewicz-Dowjat, A. J., Kelly, D. P. & Wood, A. P. (1995) Novel eubacteria able to grow on carbon disulphide. *Archives Microbiol.* **163**, 131–137.

Jordan, S. L., McDonald, I. R., Kraczkiewicz-Dowjat, A. J., Kelly, D. P., Rainey, F. A., Murrell, J. C. & Wood, A. P. (1997) Autotrophic growth on carbon disulphide is a property of novel strains of *Paracoccus denitrificans*. *Archives Microbiol.* **168**, 225–236.

Kalin, R. M., Hamilton, J. T. G., Harper, D. B., Miller, L. G., Lamb, C., Kennedy, J. T., McCauley, S. & Goldstein, A. H. (2001) Continuous flow stable isotope methods for study of the delta 13C fractionation during halomethane production and degradation. *Rapid Communications in Mass Spectrometry* **15**(5), 357–363.

Kalin, R. M. & Myles, P. (2003) A method for the treatment of carbon disulphide in soil and groundwater. *International Patent Application no. PCT/GB2003/002012*

Kalin, R. M. & Myles, P (2004) Treatment of carbon disulphide contaminated soil and groundwater using iron. In: *Abstracts In-situ and On-site Remediation of Chlorinated and Recalcitrant Compounds*. Monterey, California, USA. May 2004

Lobo, R., Revah, S. & Viveros-Garcia, T. (1999) An analysis of a trickle-bed bioreactor: carbon disulphide removal. *Biotechnology & Bioengineering*, **63**(1), 98–109.

Mackison, F. W., Sticoff, R. S. & Partridge, L. J., Jr (1981) *NIOSH/OSHA Occupational Health Guidelines for Chemical Hazards. DHHS (NIOSH) Publication no. 123*. US Government Printing Office, Washington, DC, USA.

Matheron, M. E., & Matejka, J. C. (1987) Effects of sodium tetrathiocarbonate (GY-81) on growth and sporulation of *Phytophthora citropthora* and *P. parsitica* . *Phytopathology* **77**(8), 1240.

Meidl, J. H. (1970) *Flammable Hazardous Materials*, 25–40 & 173–178. Glencoe Press, Beverly Hills, California, USA.

Morales, V., Hinojosa, A., Paez, F., Gonzalez, M., Gonzalez, O., Viveros, T. & Revah, S. (1992) Biotechnological process for the treatment of H_2S and CS_2 from a waste gas from a cellophane plant. In: *Proceedings from the Fifth International IGT Symposium on Gas, Oil, and Environmental Biotechnology, Chicago, USA*.

Moret, I., Gambaro, A., Piazza, R., Barbante, C., Andreoli, C., Corami, F. & Scarponi, G. (2000) The seasonal variations of dimethyl sulphide and carbon disulphide in surface waters of the Venice lagoon. *Marine Chemistry* **71**, 283–295

Muezzinoglu, A. (2003) A study of volatile organic sulfur emissions causing urban odors. *Chemosphere* **51**, 245–252

Nederlands Instituut voor Arbeidsomstandigheden (NIA) (1988) *Chemiekaarten,* 4th edn. Stuurgroep Chemiekaarten, Samson Uitgeverij bv., Alphen aan den Rijn: p. 1038.

NPL Site Narrative for Avtex Fibers, Inc. (2001) http://www.epa.gov/oerrpage/superfund/sites/npl/nar362.htm

NPL Site Narrative for Halby Chemical Co. (2001) http://www.epa.gov/oerrpage/superfund/sites/npl/nar259.htm

NPL Site Narrative for Stauffer Chemical Co. (LeMoyne Plant) (2001) http://www.epa.gov/oerrpage/superfund/sites/npl/nar387.htm

Padden, N. A. (1997) Microbial degradation of organic sulphur pollutants. PhD Thesis. Available at: http://members.tripod.com/~napadden.htm

Plas, C., Wimmer, K., Holubar, P., Mattanovich, D., Danner, H., Jelinek, E., Harant, H. & Braun, R. (1993) Degradation of carbon disulphide by a *Thiobacillus* isolate. *Applied Microbiol. Biotechnology* **38**, 820–823.

Revah, S., Acosta, M., Hugler, W., Trinidad, R., Avila, C., Estrada, I. & Hinojosa, A. (1995) Air biodesulphurization from viscose plants: Carbon disulphide elimination. In: *Proc. 1995 Conference on Biofiltration, Los* Angeles, 181–187. University of Southern California, California, USA.

Rhodes, C., Riddel, S. A., West, J., Williams, P. & Hutchings, G. J. (2000) The Low-temperature hydrolysis of carbonyl sulphide and carbon disulphide. *Catalysis Today* **59**(3–4), 443–464.

Scholler, C., Molin, S. & Wilkins, K (1997) Volatile metabolites from some gram-negative bacteria. *Chemosphere*, **35**(7), 1487–1495

Seidell, A. & Linke, W. F. (1952) *Solubilities of Inorganic and Organic Compounds*. D. Van Nostrand, New York 101, USA.

Smith, N. A. & Kelly, D. P. (1988) Oxidation of carbon disulphide as the sole source of energy for the autotrophic growth of *Thiobacillus thioparus* strain TK-M. *J. Genetic Microbiol.* **134**, 3041–3048.

Sorokin, D. Y., Tourova, T. P., Lysenko, A. M. & Kuenen, J. G. (2001) Microbial thiocyanate utilisation under highly alkaline conditions. *Appl. Environ. Microbiol.* **67**(2), 528–538

Tratnyek, P. G. (1996) Putting corrosion to use: Remediation of contaminated groundwater with zero-valent metals. *Chemistry & Industry* **13**, 499–503.

US EPA (1998a) *In situ* remediation technology: *in situ* chemical oxidation. office of solid waste and emergency response. *Washington, D.C. EPA 542-R-98-008*.

US EPA (Feb. 2001) Superfund sites index page: http://www.epa.gov/superfund/sites/index.htm

US National Technical Information Service (1989) *Industrial Exposure and Control Technologies for OSHA Regulated Hazardous Substances* Vol. I, 366–369. Springfield, Virginia, USA.

US Patent 5 536 898

US Patent 6 283 675

Vaccari, A., Saba, P., Mocci, I. & Ruiu, S. (1999) Dithiocarbamate pesticides affect glutamate transport in brain synaptic vesicles. *J. Pharmacology and Experimental Therapeutics* **288**, 1–5.

Van Groenestijn, J. W. & Hesselink, P. G. M. (1993) Biotechniques for air pollution control. *Biodegradation*, **4**, 283–301.

Watts, S. F. (2000) The mass budgets of carbonyl sulphide, dimethyl sulphide, carbon disulphide and hydrogen sulphide. *Atmos. Environ.* **34**, 761–779.

Weber, J. (1999) Volatile hydride and methyl compounds of selected elements formed in the marine environment. *Marine Chemistry* **65**, 67–75.

Wijnands, F. G. & van Asperen, P. (1999) Milieubelasting verminderen door gerichte middelenkeuse. *PAV Bulletin Akkerbouw*, Juni, 28–37.

Young, D. G. & Green, J. A., II (1988) Stabilized thiocarbonate solutions and the use thereof for controlling soil pests. *US Patent 4 726 144*.

Nitrate reduction without ammonium release at circum-neutral pH

SEUNGHAK LEE, KWANGHUN LEE, SUNGSU RHEE, AHRA LEE & JUNBOUM PARK
School of Civil, Urban & Geosystem Engineering, Seoul National University, Seoul, 151-742, Korea
junbpark@snu.ac.kr

Abstract Nitrate reduction by zero valent iron (ZVI) has been previously studied, but the proper treatment for the by-product of ammonium has not been reported. This study is focused on simultaneous removal of nitrate and its by-product of ammonium, with the ZanF (zeolite anchored Fe), a product derived from zeolite modified by Fe(II) chloride followed by reduction with sodium borohydride. Batch experiments were performed without a buffer at different pH condition with ZVI or ZanF. ZanF was found to be more appropriate in PRBs than ZVI for treating nitrate by reducing nitrate without pH control at initial circum-neutral pH and sorbing the ammonium by-product simultaneously.

Key words ammonium; nitrate; simultaneous removal; ZanF

INTRODUCTION

As mentioned by Cheng *et al.* (1997), the main disadvantages of nitrate reduction using ZVI are the requirement of extremely low pH, or pH buffering, and the production of ammonium. When applying ZVI in PRBs for nitrate removal, these disadvantages are even more critical. It would be both undesirable and difficult to reduce the pH of an aquifer to such a low value. Moreover, in Korea, the acceptable level of ammonium in drinking water, 0.5 mg-N l^{-1}, is much lower than that of nitrate (10 mg-N l^{-1}).

As an alternative to ZVI, the current study focuses on developing new materials appropriate in PRBs, which can reduce nitrate at circum-neutral pH without the necessity of a buffer, and properly remove the ammonium by-products. The optimal conditions for making ZanF were determined, and the performance of ZanF for nitrate removal was verified by conducting a series of batch experiments.

EXPERIMENTAL SECTION

Preparation of ZanF

Five hundred grams of zeolite were stirred with 5 l of 350 mM $FeCl_2$ solution in an acrylic chamber for 24 h to introduce Fe(II) in zeolite. After stirring, the solution was decanted, and the solids were washed with deionized water to remove excess $FeCl_2$. Then the solids were stirred in 5 l of 23 mM sodium borohydride solution ($NaBH_4$,

98%, Sigma, USA) for 1 h to reduce Fe(II) in the zeolite to Fe(0). The borohydride-reduced zeolite was washed three times with deionized water, and dried in the vacuum oven at 90°C for 2–3 h, and stored under anaerobic conditions. The materials made by the above process were referred to as ZanF.

Nitrate removal with iron or ZanF at different pH

By dissolving KNO_3 (assay 98%, Shinyo Pure Chemicals Co., Japan) in deionized water the initial pH of solution was varied from 3 to 7. Each solution was purged with argon gas for 10–15 min to remove the dissolved oxygen. After purging, solutions were filled into a series of 26 ml amber glass bottles containing 3 g of iron filing (about 40-mesh, Fisher Scientific, USA) or ZanF leaving no headspace. The bottle was mixed on a vertical rotary shaker at 36 rpm for 36 h at ambient temperature. At pre-determined time intervals, each bottle was taken, and the solution was analysed for pH, and residual concentration of nitrate and ammonium. After reaction, the ZanF was washed with deionized water, and then treated with 1 M KCl solution to measure the ammonium sorbed.

RESULTS AND DISCUSSION

Nitrate reduction at different pH conditions with ZVI

Figures 1 and 2 show the nitrate reduction with ZVI under different initial pH, and the ammonium production during the reaction, respectively. The initial pH effect was negligible, even though previous research reported that nitrate reduction could be enhanced in acidic conditions (Cheng *et al.*, 1997; Huang *et al.*, 1998). From the mass balance analysis (not shown here), it was found that the ratio of nitrogen in the residual solution (NO_3-N + NH_4^+-N) to initial nitrate was maintained at around unity throughout the reaction. Therefore, it was confirmed that nitrate was transformed mainly to ammonium. The results of nitrate reduction with ZVI suggested that when applied in PRBs, ZVI could be limited by: (a) the low nitrate reduction efficiency under typical pH conditions in an aquifer, and (b) the ammonium production during the reduction reaction. As mentioned before, this ammonium by-product is also considered as a contaminant at concentrations >0.5 mg-N l^{-1} in drinking water in Korea.

Nitrate reduction at different pH conditions using ZanF

Also in nitrate reduction by ZanF, the initial pH effect was not significant. As shown in Fig. 3, more than 50% of nitrate was removed after 36 h. Importantly, ZanF showed significant nitrate removal even under near-neutral pH conditions of 6.2 and 6.9 (5.33 mg N g^{-1}), whereas the extent of nitrate removal by ZVI was limited to only about 13% (1.39 mg-N g^{-1}). With regard to by-products, neither ammonium nor nitrite were detected in the residual solution during the reaction. It was presumed that

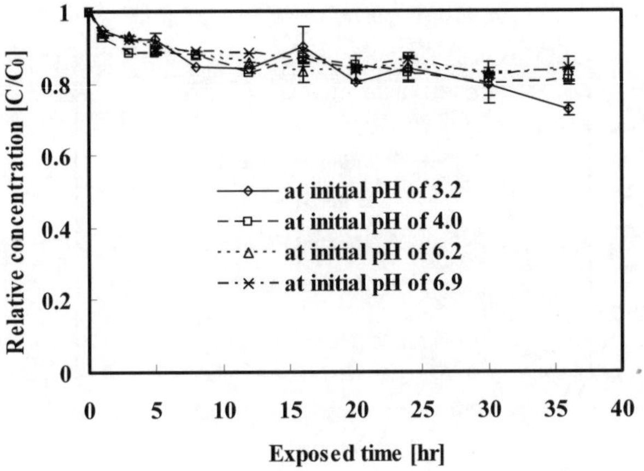

Fig. 1 Nitrate reduction with ZVI.

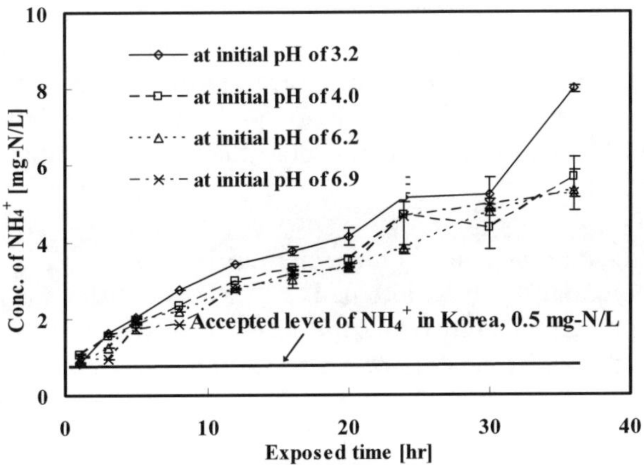

Fig. 2 NH_4^+-N production in ZVI-system.

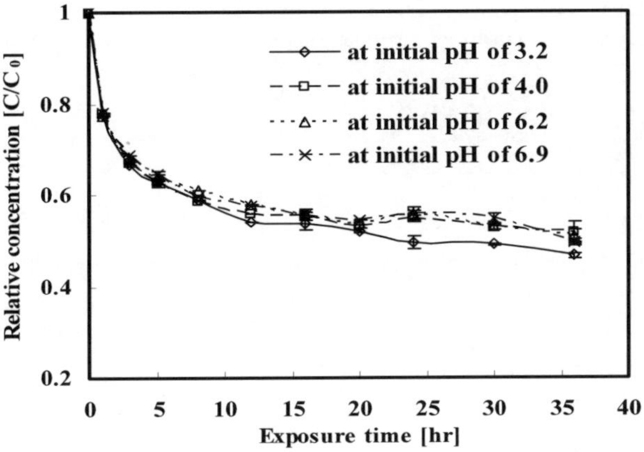

Fig. 3 Nitrate reduction with ZanF.

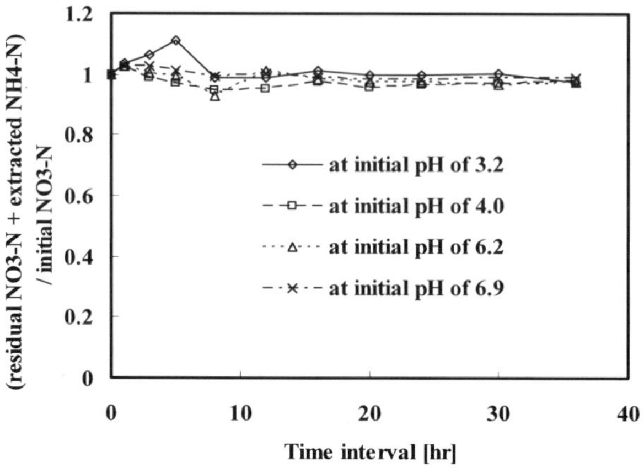

Fig. 4 Mass balancee on N-species in residual solution.

ammonium might be rapidly sorbed in ZanF through ion exchange reaction. From the results of KCl extraction, the amount of ammonium sorbed in ZanF was found to increase as the nitrate reduction proceeded, indicating that nitrate was transformed to ammonium, and that it was sorbed in ZanF. A mass balance on N-species (residual NO_3-N and extracted NH_4-N) accounted for more than 98% of initial nitrate (Fig. 4), indicating that ammonium produced from nitrate reduction was completely removed by ZanF.

These results indicate that ZanF is more appropriate than ZVI in treating nitrate, and may have the potential to be an alternative to ZVI in PRBs for nitrate removal. However, to confirm the applicability of ZanF in PRBs, a series of further issues have to be addressed, including the longevity of materials and economic analysis in practical installation.

REFERENCES

Cheng, I. F., Muftikian, R., Fernando, Q. & Korte, N. (1997) Reduction of nitrate to ammonia by zero-valent iron. *Chemosphere* **35**(11), 2689–2695.

Huang, C. P., Wang, H. W. & Chiu, P. C. (1998) Nitrate reduction by metallic iron. *Water Res.* **32**(8), 2257–2264.

Assessment of different carbon sources and delivery techniques to promote an *in situ* reactive zone for bioprecipitation of metals in groundwater

BRADLEY M. PATTERSON[1], YAMIN MA[1,2], MICHELLE E. GRASSI[1,2], BLAIR S. ROBERTSON[1], GREG B. DAVIS[1], MARK LIPMAN[3], STUART RHODES[3], ALLAN J. MCKINLEY[2] & ANDREW W. RATE[2]

1 *CSIRO Land and Water, Private Bag No. 5, Wembley, Western Australia 6913, Australia*
bradley.patterson@csiro.au

2 *The University of Western Australia, Nedlands, Western Australia 6009, Australia*

3 *Rio Tinto Technical Services, Milsons Point, New South Wales 2061, Australia*

Abstract Large-scale soil column experiments were undertaken using low metal-sorbing soil to assess the efficiency of different carbon sources to promote and maintain a sulphate-reducing *in situ* reactive zone for bioprecipitation of metals in groundwater. Carbon sources were delivered by either direct injection (molasses), diffusion delivery through polymers (ethanol) or as a non-aqueous phase liquid (emulsified vegetable oil). All three carbon sources promoted denitrification with denitrification half lives between 1.3 and 2.1 days. However, only molasses and ethanol promoted sulphate reduction, with average sulphate-removal efficiencies of 53% and 95%, and sulphate reduction half lives of 8.3 and 1.2 days, respectively. Based on this information, ethanol was a more efficient carbon source in promoting and maintaining sulphate reduction within the column compared to molasses. Under these experimental conditions, non-aqueous phase vegetable oil amendment did not prove suitable as a carbon source to promote sulphate reduction. Substantial removal of zinc and copper (>98%) was observed in groundwater for both the ethanol-amended and molasses-amended columns. Zinc and copper were removed from the groundwater as it flowed past the sulphate-reducing regions of the columns, suggesting that the metals were deposited on the soil, possibly as zinc and copper sulphides. The longevity of sulphate reduction after ethanol and molasses delivery had ceased was in the order of two weeks or less, suggesting that in a field application, relatively continuous carbon addition would be required to maintain sulphate reduction for long-term treatment of metal-contaminated groundwater.

Key words bioprecipitation; groundwater; *in situ* reactive zone; metals; sulphate reduction

INTRODUCTION

Recently, there has been a growing interest in bioprecipitation as a metal-remediation technique. This involves the precipitation of metal sulphides through the biostimulation of sulphate-reducing bacteria (SRB) (Benner *et al.*, 1999; Blowes *et al.*, 2000; Foucher *et al.*, 2001; Waybrant *et al.*, 2002; Grassi *et al.*, 2003). Under favourable conditions, SRB catalyse the oxidation of organic carbon: (a) via sulphate-reduction to produce hydrogen sulphide that reacts with metal ions; (b) present in solution to produce stable metal-sulphides of low solubility.

$$2CH_3CH_2OH + 3SO_4^{2-} + 3H^+ \rightarrow 3HCO_3^- + 3H_2S + 3H_2O + CO_2 \quad (1)$$

$$H_2S + Metal^{2+} \rightarrow MetalS(s) + 2H^+ \quad (2)$$

Application of bioprecipitation as an *in situ* reactive zone (ISRZ) for the treatment of metal-contaminated groundwater should be feasible, however, the selection and delivery of a non-pollutant carbon substrate to promote bioprecipitation has yet to be evaluated fully. This paper details with the results of large-scale soil column investigations to assess the efficiency of different carbon sources (molasses, ethanol and emulsified vegetable oil) to promote and maintain sulphate reduction and metal sulphide precipitation.

METHODS

Column construction

Three carbon-amended columns and a control column were constructed to assess the efficiency of different carbon sources to promote and maintain sulphate reduction and metal precipitation. Each column (150 mm i.d., 1.90 m long) was fitted with 13 water-sampling ports along the length of each column. Gravity volumetric flow through each column was restricted using a peristaltic pump on the effluent line to give a flow velocity through the sediments of ~25 m $year^{-1}$. Experiments were run at room temperature, which was controlled coarsely at 22°C.

The columns were filled with a highly bleached, low organic (200 mg kg^{-1}) and iron (13 mg kg^{-1}) content sand collected from the Swan Coastal Plain, Perth Western Australia. Groundwater from the site contained sulphate (~30 mg l^{-1} SO_4-S), nitrate (~4.5 mg l^{-1} NO_3-N), dissolved oxygen (2.1 mg l^{-1}) and dissolved organic carbon (DOC) content (~4 mg l^{-1} C). The groundwater was spiked with a concentrated metal chloride solution to give a final concentration of 12 mg l^{-1} for Zn^{2+} and 4.0 mg l^{-1} for Cu^{2+}. The groundwater pH was maintained at 5.5. The soil and groundwater in the columns were used without bacterial amendment (no bioaugmentation).

Carbon delivery

A sterilized concentrated molasses solution (17 g l^{-1}) was injected (at 10 ml day^{-1}) into the molasses-amended column through six Teflon-coated porous tubes with holes (0.5-mm diameter) every 1 cm apart, giving a total of 64 potential delivery locations. The Teflon-coated tubes were emplaced horizontally within the circumference of the column 0.5 m from the base of the column. Dilution of the concentrated molasses solution with the groundwater flowing past the delivery location gave a groundwater molasses concentration of ~1 g l^{-1}.

Ethanol was delivered to the ethanol-amended column, via diffusion through a polymer mat located 0.5 m above the base of the column, and emplaced horizontally within the circumference of the column. The polymer mat consisted of 5.0 m of poly(dimethylsiloxane) tubing (3.0 mm o.d., 2.0 mm i.d.) woven into a 13.5 cm diameter flexible plastic support frame, and contained a coiled stainless steel wire to prevent collapsing and twisting of the tubing. For ethanol delivery, a concentrated ethanol solution (between 5 and 10 g l^{-1}) was pumped through the lumen of the polymer tube at a rate of 2 ml min^{-1} to give an influent ethanol groundwater concentration of 0.25 to 0.8 g l^{-1}.

A once-off emulsified vegetable oil solution (consisting of 16.9 g of vegetable oil (Crisco's soybean oil), 34 g of distilled water and 0.85 g of Tween 85 (40, 975-8 *n,n*-dimethyl-phenlyenediamine oxalate)) was injected into the oil-amended column through four delivery ports located horizontally within the circumference of the column 0.5 m from the base of the column. Using a long stainless steel needle, the emulsion was delivered in ~1 g aliquots at multiple horizontal locations 0.5 m from the base of the column.

RESULTS AND DISCUSSION

Establishment of reducing conditions

After the different carbon sources were added to the columns, effluent groundwater in the molasses-amended column showed a substantial reduction in Eh to –200 mV (Fig. 1), after ~4 weeks of molasses delivery (Table 1). For the ethanol-amended column, partial blockage of the influent groundwater for the first 5 weeks, resulted in inconsistent (stop/start flow) groundwater flow through this column. Once this problem was rectified, consistent groundwater flow through this column was observed. Approximately 2 weeks after the groundwater flow problem was rectified, Eh values decreased to approximately –200 mV, and remained low until ethanol delivery was

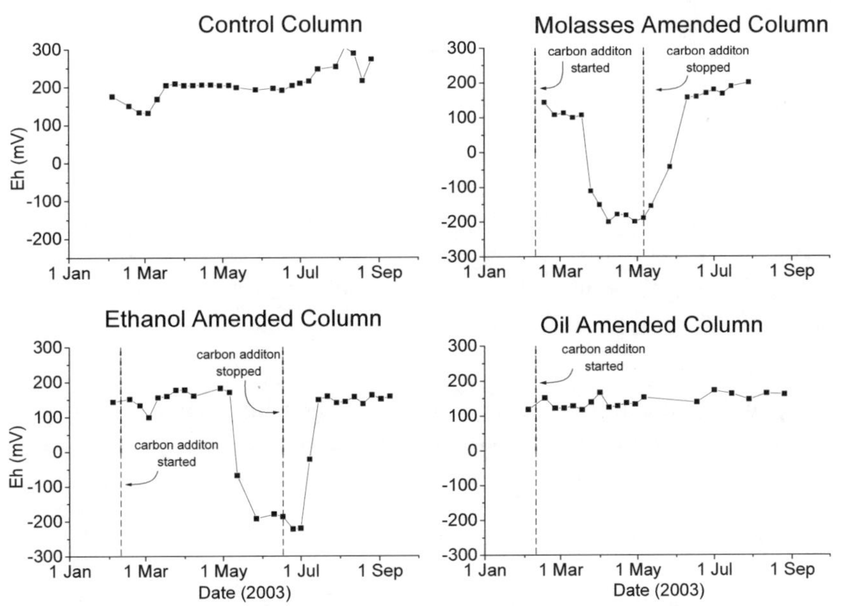

Fig. 1 Eh values of effluent groundwater from each column.

Table 1 Denitrification, sulphate reduction and metal removal parameter for different columns.

Parameter	Amendment			
	Ethanol	Molasses	Oil	Control
Eh (mV)	–200	–200	100–200	100–300
Denitrification				
Lag-time before initiation of denitrification (days)	15	15	29	>200
Average denitrification half life at steady-state (days)	1.3 (±0.5)	1.4 (±1.4)	2.1 (±1.4)	>15
Average nitrate removed at steady-state (%)	100	100	83	<5
Sulphate reduction				
Lag-time before initiation of sulphate reduction (days)	99 (±13)	43 (±6)	>200	>200
Average sulphate reduction half-life at steady-state (days)	1.2 (±0.6)	8.3 (±3.4)	>15	>15
Average sulphate removed at steady-state (%)	95	53	<5	<5
Metal removal				
Zinc removal (effluent to influent concentrations) (%)	99.9	99.3	10	<10
Copper removal (effluent to influent concentrations) (%)	99.7	98.1	86	<10

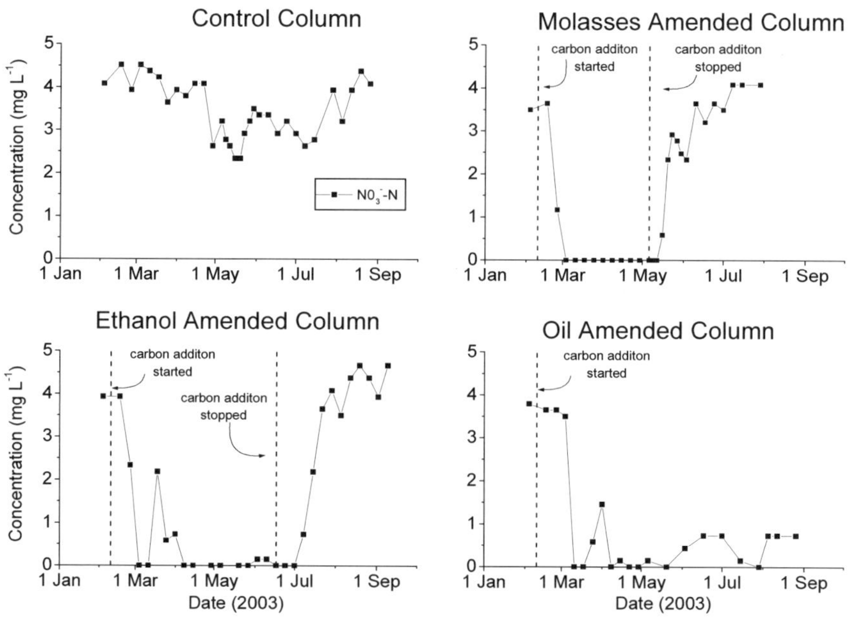

Fig. 2 Nitrate concentrations of effluent groundwater from each column.

stopped. In the control column, Eh values fluctuated between 100 and 300 mV, while Eh values in the oil-amended column fluctuated between 100 and 200 mV.

Denitrification

Denitrification within the carbon-amended columns occurred relatively rapidly once the different carbon sources were added to the columns (Fig. 2). Lag times before denitrification commenced were 15 days for the molasses and ethanol-amended columns, and 29 days for the oil-amended column, respectively. Consistent and complete denitrification was observed in the molasses-amended column during molasses delivery. For the ethanol-amended column, consistent and complete denitrification occurred once the partial blockage problem in this column was rectified. The oil-amended column also showed denitrification once the oil was added. However, denitrification was inconsistent within this column.

After denitrification rates stabilized, average nitrate mass removal and first order denitrification half lives for each of the columns were determined (Table 1). These results suggest that both molasses and ethanol provided a similar level of efficiency in promoting and maintaining denitrification, while the oil-amendment was less efficient. However, denitrification continued in the oil-amended column until this column was decommissioned, a total of ~28 weeks. Therefore, the emulsified oil treatment seemed to be able to sustain denitrification over a relatively longer time, although the nitrate removal was less effective than the other two treatments.

Sulphate reduction

For the molasses-amended column, sulphate reduction (decrease in sulphate concentration and increase in sulphide concentration) (Fig. 3), commenced ~6 weeks after molasses delivery began. For the ethanol-amended column, sulphate reduction commenced ~14 weeks after ethanol delivery began. (~5 weeks after the groundwater flow problem was rectified). No sulphate reduction was observed in the oil-amended column.

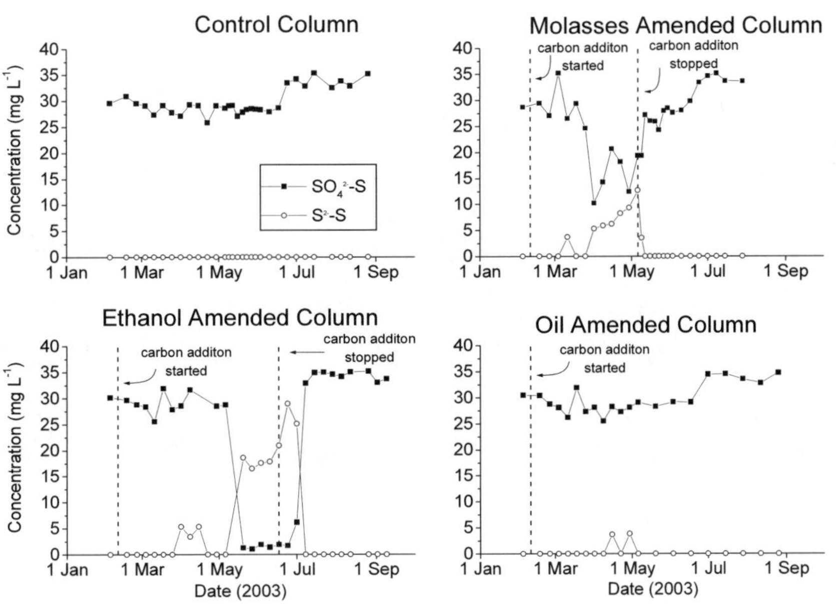

Fig. 3 Sulphate and sulphide concentrations of effluent groundwater from each column.

The average mass removal and first order sulphate reduction half-lives estimated for each column is given in Table 1. For the molasses-amended column, an average sulphate removal efficiency of 53% (compared to influent concentrations), and a sulphate reduction half-life of 8.3 days were estimated. In contrast, sulphate reduction in the ethanol-amended column was greater with an average of 95% sulphate removed with a sulphate reduction half-life of 1.2 days.

These results suggest that ethanol provides a potentially more efficient carbon source in promoting and maintaining sulphate reduction than molasses. In both the ethanol and molasses amended columns, sulphate reduction was not carbon limited (complete carbon removal did not occur). The lower sulphate reduction half life in the ethanol-amended column may be that ethanol is used directly as an electron donor by SRB, while molasses (sucrose) needs to be fermented into low molecular weight monomers before most SRB can utilize this carbon source. Under these experimental conditions, the oil amendment was not a suitable carbon source to promote sulphate reduction. The inability for the SRB to utilize the vegetable oil and induce sulphate reduction may be due to the higher energy demand on the SRB to break down and assimilate the large molecules of oil (Coyne, 1999). Also, the groundwater sulphate concentrations are relatively low, which may limit sulphate reduction.

Groundwater metal removal

Within 2–4 weeks after the carbon sources were added (Fig. 4), dissolved zinc and copper concentrations in column effluent groundwaters of the three carbon-amended

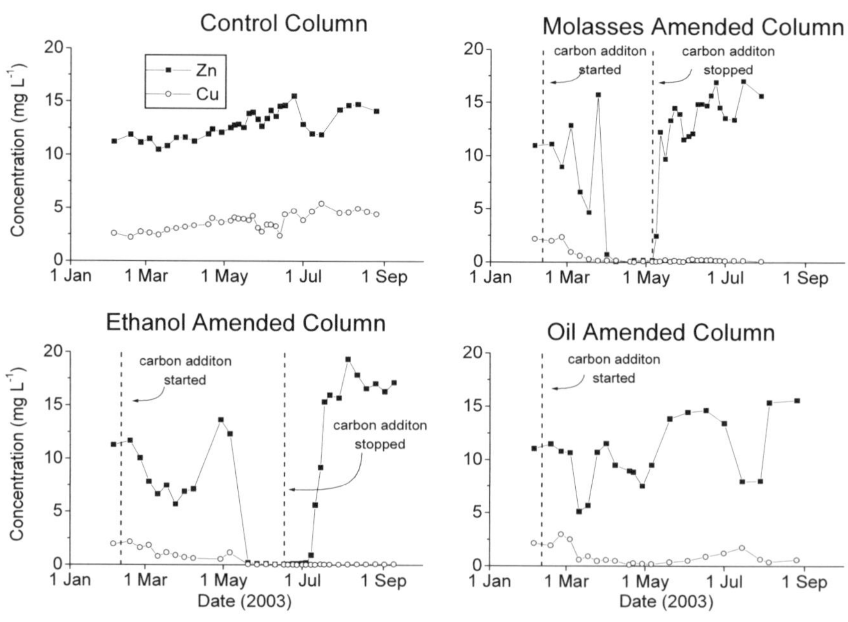

Fig. 4 Zinc and copper concentrations of effluent groundwater from each column.

columns were reduced. This initial reduction in dissolved metal concentrations did not correspond to the time that sulphate reduction was observed. Also, complete metal removal did not occur during this time. These data indicated that a non-sulphate reduction mechanism such as biosorption (Chatterjee *et al.*, 1996) or the formation of metal carbonates (McGregor *et al.*, 2002) was possibly responsible for the reduction in dissolved metal concentrations at this time.

Once near steady-state sulphate-reduction was achieved, >98% of dissolved zinc and copper was removed (compared to influent concentrations) in this zone of sulphate reduction, for both the ethanol-amended and molasses-amended columns (Fig. 4). In general, when sulphate reduction was near steady-state, both the ethanol and molasses-amended columns achieved similar amounts of metal removal. Hence, it may be concluded that for metal remediation purposes, both the ethanol and molasses treatments produced comparable results in attenuating similar amounts of metals.

Cessation of carbon addition

After carbon-addition ceased, Eh values increased, and denitrification and sulphate reduction ceased in the molasses and ethanol amended columns (Figs 1, 2, 3). Zinc and copper concentrations were observed to increase in locations in columns that previously showed sulphate reduction. The migration of zinc and copper through the columns were consistent with previously observed retardation coefficients. In the molasses and ethanol-amended columns, effluent zinc (retardation coefficient of 1.5) concentrations increased and stabilized to near influent concentrations within 2–4 weeks. Effluent copper (retardation coefficient of 3.3) concentrations also increased, however concentrations did not stabilize to near influent concentrations within the 12-week monitoring period after carbon addition ceased.

CONCLUSIONS

Large-scale column experiments have shown that the establishment of a sulphate reducing ISRZ for the bioprecipitation of metal-contaminated groundwater is feasible, using either direct aqueous addition of molasses or diffusion delivery of ethanol. A lower sulphate reduction half life was achieved using ethanol compared to molasses, possibly due to the more direct carbon utilization of ethanol compared to molasses. However, continuous carbon source delivery is required to maintain long-term treatment. The once-off addition of emulsified oil was unable to promote sulphate reduction and metal bioprecipitation.

Acknowledgement The authors wish to thank N. Innes from CSIRO Land and Water for his technical assistance with this work. The authors are also grateful to Rio Tinto for funding this work.

REFERENCES

Benner, S. G., Blowes, D. W., Gould, W. D., Herbert, W. B., Jr. & Ptacek, C. J. (1999) Geochemistry of a permeable reactive barrier for metals and acid mine drainage. *Environ. Sci. Technol.* **33**, 2793–2799.

Blowes, D. W., Ptacek, C. J., Benner, S. G., McRae, C. W. T., Bennett, T. A. & Puls, R. W. (2000) Treatment of inorganic contaminants using permeable reactive barriers. *J. Contaminant Hydrol.* **45**, 123–137.

Chatterjee, S., Asthana, R. K., Tripathi, A. K. & Singh S. P. (1996) Metal removal by selected sorbents. *Process Biochem.* **31**(5), 457–462.

Coyne, M. (1999) *Soil Microbiology: An Exploratory Approach.* Delmar Publishers, Tennessee, USA.

Foucher, S., Battaglia-Brunet, F., Ingatiadis, I. & Morin, D. (2001), Treatment by sulfate-reducing bacteria of Chessy acid-mine drainage and metals recovery. *Chem. Engng Sci.* **56**, 1639–1645.

Grassi, M. E., Patterson, B. M., McKinley, A. J. & Davis, G. B. (2003) Remediation of metal-contaminated groundwater using an innovative carbon source delivery system. In: *Proc. Ozwater 2003, Perth, Western Australia, April 2003.* oz029.

McGregor, R., Benner, S., Ludwig, R., Blowes, D. & Ptacek, C. (2002) Sulfate reduction permeable reactive barriers to treat acidity, cadmium, copper, nickel and zinc: two case studies. In: *Handbook of Groundwater Remediation Using Permeable Reactive Barriers* (ed. by D. L. Naftz, C. C.Fuller, S. J. Morrison & J. A. Davis), 495–522. Elsevier Science, USA.

Waybrant, K. R., Ptacek, C. J. & Blowes, D. W. (2002) Treatment of acid mine drainage using permeable reactive barriers: column experiments. *Environ. Sci. Technol.* **36**, 1349–1356.

Synthetic zeolites as sorbent material for PRBs at industrially contaminated sites

R. VIGNOLA[1], G. GRILLO[1], R. SISTO[1], G. CAPOTORTI[2], P. CESTI[2] & M. MOLINARI[3]

1 *EniTecnologie S.p.A. via E. Ramarini 32, I-00016 Monterotondo, Rome, Italy*
rvignola@mail.enitecnologie.eni.it

2 *Polimeri Europa S.p.A, Istituto G. Donegani, via G. Fauser 4, I-28100 Novara, Italy*

3 *Eni S.p.A, Divisione Refining & Marketing, via Laurentina 449, I-00142 Rome, Italy*

Abstract An innovative system for the treatment of groundwater contaminated by different classes of organic pollutants is proposed: hydrocarbons, chlorinated compounds, such as 1,2 dichloroethane (1,2 DCA), or methyl-ter-butyl-ether (MtBE). The use of hydrophobic zeolites (HZ) overcomes the problems associated with the treatment of these compounds, whose remediation is very difficult by currently available technologies based on activated carbon (GAC) or zerovalent iron (ZVI). In fact, the selectivity of HZ is based on chemical-physic affinities of adsorbate to adsorbent. A wide range of pollutants related to oil industry activities has been examined: chlorinated ethanes, ethenes and benzenes, mono and polynuclear aromatic hydrocarbons. The results showed good to excellent sorption capacity and velocity. The complete remediation of two cases is demonstrated: the first is relative to groundwater containing aromatic hydrocarbons in the presence of MtBE, the second to groundwater contaminated by 1,2 DCA and chlorinated alkenes. It has been shown that sorption capacity of zeolites was not inhibited by the presence of high concentrations of ions commonly present in groundwater, even in coastal sites. The experimental activities have been carried out in laboratory simulation apparatus.

Key words 1,2 DCA; groundwater contamination; MtBE; zeolite

INTRODUCTION

One of the most formidable challenges posed by increasing stringent regulation of water quality is the search for efficient and economical treatment strategies for groundwater contaminated by different classes of organic pollutants, especially those whose remediation is very difficult to obtain by means of currently available technologies, e.g. 1,2-dichloroethane (1,2-DCA) or chlorobenzenes (CB_S).

The commonly used technology for a wide range of contaminants, such as chlorinated solvents and hydrocarbons, is adsorption on granulated activated carbon (GAC). However, GAC presents several disadvantages due to the physico-chemical characteristics of its surface: pore plugging, interactions with humic substances and inorganic ions, adverse effect of pH on the adsorption of organics (Fajula *et al.*, 1994). Additionally, GAC has been shown to be slightly effective in treatment of water containing very soluble compounds, such as oxygenated organics, or low molecular weight compounds, such as vinyl chloride (VC) (Watson, 1996; Liang *et al.*, 1999; EDAG, 2001).

Recently, the use of permeable reactive barriers (PRB) based on ZVI as the reactive material has been developed. However, this material appears to be ineffective on irreducible compounds such as 1,2-DCA and CB_S as well as hydrocarbons (Schad *et al.*, 2000). Further efforts have been placed on finding innovative reactive systems either following the reductive approach by introducing other metals, such as Pd, Ni, or using more effective sorbents, such as Amberlites™ (Scott & Powers, 2000).

This paper discusses the use of HZ as an innovative sorbent material suitable for the *in situ* abatement of contaminants in groundwater through PRB.

RATIONAL

Industrial megasites in the petrochemical industry and refinery facilities, are frequently situated on the coast. Such a situation can determine a complex contamination of the groundwater, in which a mixture of organic compounds, including chlorinated alkanes/alkenes and hydrocarbons, has often been found. Besides, the remediation can be complicated by the occurrence of particular compounds that are difficult to treat with commercial technologies such as 1,2-DCA and MtBE. Moreover, the presence of high salt concentrations constitutes a further difficulty.

Synthetic HZ can provide an effective solution due to their high selectivity, adsorption rapidity, good adsorption capacity and easy regeneration. Their selectivity is based on physical-chemical affinities of adsorbate and adsorbent. Properties of HZ determine no interactions with polar interfering substances, while attraction of hydrophobic organic contaminants is promoted: hence higher performances are even expected in coastal industrial sites.

Adsorption occurs if pore diameters match the dimensions of the contaminant molecules; only molecules whose sizes and polarities are consistent with the zeolite characteristics can enter the channel system and interact with its framework. Molecules are so differentiated on the basis of size and polarity. Zeolites used in this work have been specifically selected as a function of organic pollutants to be removed from groundwater.

RESULTS AND DISCUSSION

PRBs have been simulated in the laboratory by means of columns (7 cm diameter and 40 cm length) filled with suitable zeolites, fed with groundwater sampled from industrially contaminated aquifers, at a flow rate of 30–100 cm day^{-1} for 4–6 months. Two significant examples are illustrated below (Figs 1 and 2).

The first case relates to groundwater contaminated by 37 mg l^{-1} 1,2-DCA and other chlorinated compounds: 3.3 mg l^{-1} vinyl chloride (VC), 3.3 mg l^{-1} 1,2-cis-dichloroethylene (1,2-cDCE), 2.0 mg l^{-1} trichloroethylene (TCE), 3.5 mg l^{-1} tetrachloroethylene (PCE), 1.6 mg l^{-1} hexachloroetane (HCA), 1.4 mg l^{-1} hexachlorobutadiene (HCBd). Salinity is higher than 5 g l^{-1} as NaCl.

The second one concerns groundwater heavily contaminated by MtBE (17.9 mg l^{-1}) in the presence of aromatic hydrocarbons: mainly benzene (2.8 mg l^{-1}), toluene

(0.5 mg l^{-1}), ethylbenzene (1.5 mg l^{-1}), *o*-xylene (1.3 mg l^{-1}), *p*-xylene (2.9 mg l^{-1}). The salinity is 15 g l^{-1} as NaCl. Figure 1 reports GC-MS profiles at the inlet (Fig. 1(a)) and outlet (Fig. 1(b)) of the simulation apparatus, respectively.

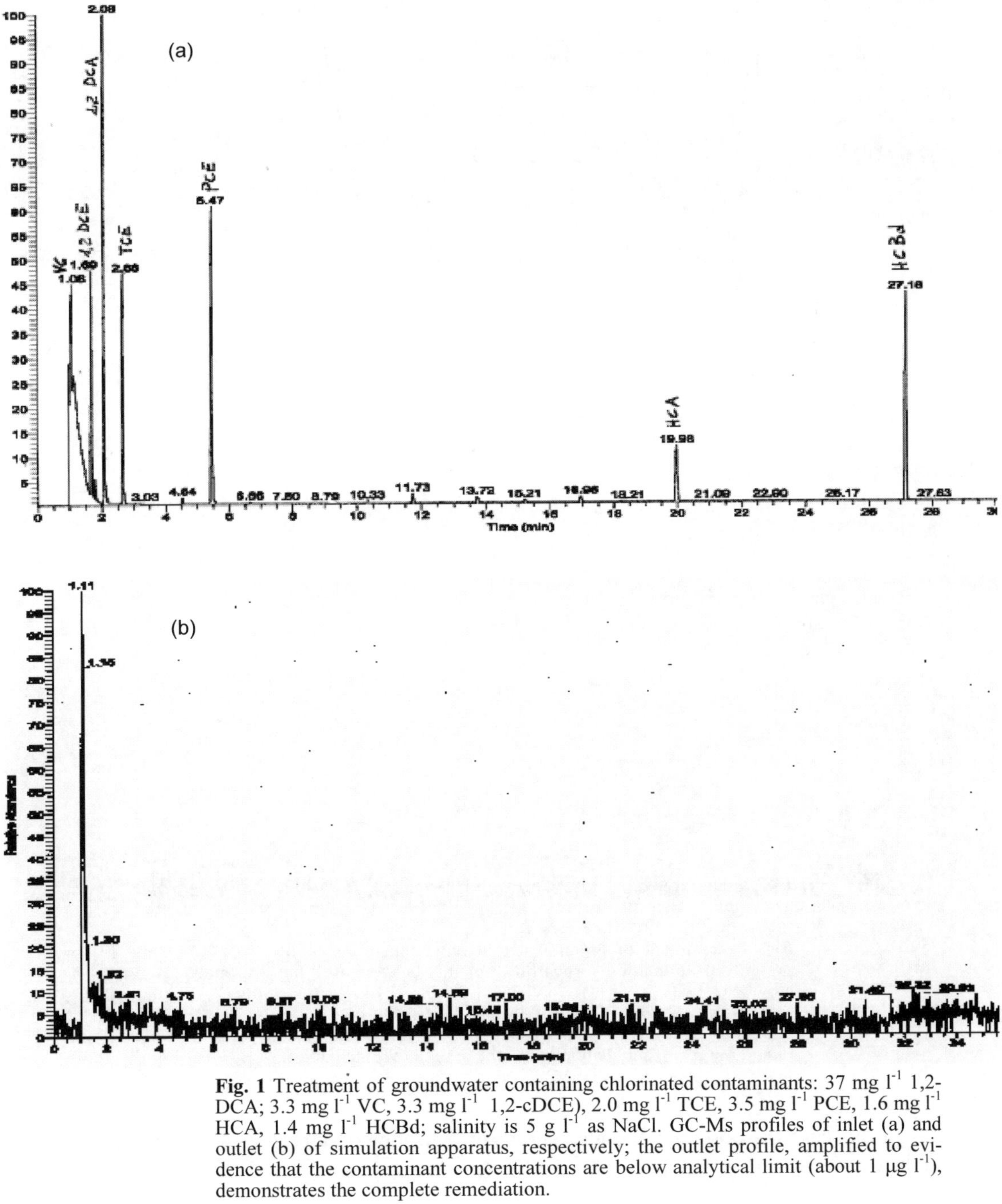

Fig. 1 Treatment of groundwater containing chlorinated contaminants: 37 mg l^{-1} 1,2-DCA; 3.3 mg l^{-1} VC, 3.3 mg l^{-1} 1,2-cDCE), 2.0 mg l^{-1} TCE, 3.5 mg l^{-1} PCE, 1.6 mg l^{-1} HCA, 1.4 mg l^{-1} HCBd; salinity is 5 g l^{-1} as NaCl. GC-Ms profiles of inlet (a) and outlet (b) of simulation apparatus, respectively; the outlet profile, amplified to evidence that the contaminant concentrations are below analytical limit (about 1 μg l^{-1}), demonstrates the complete remediation.

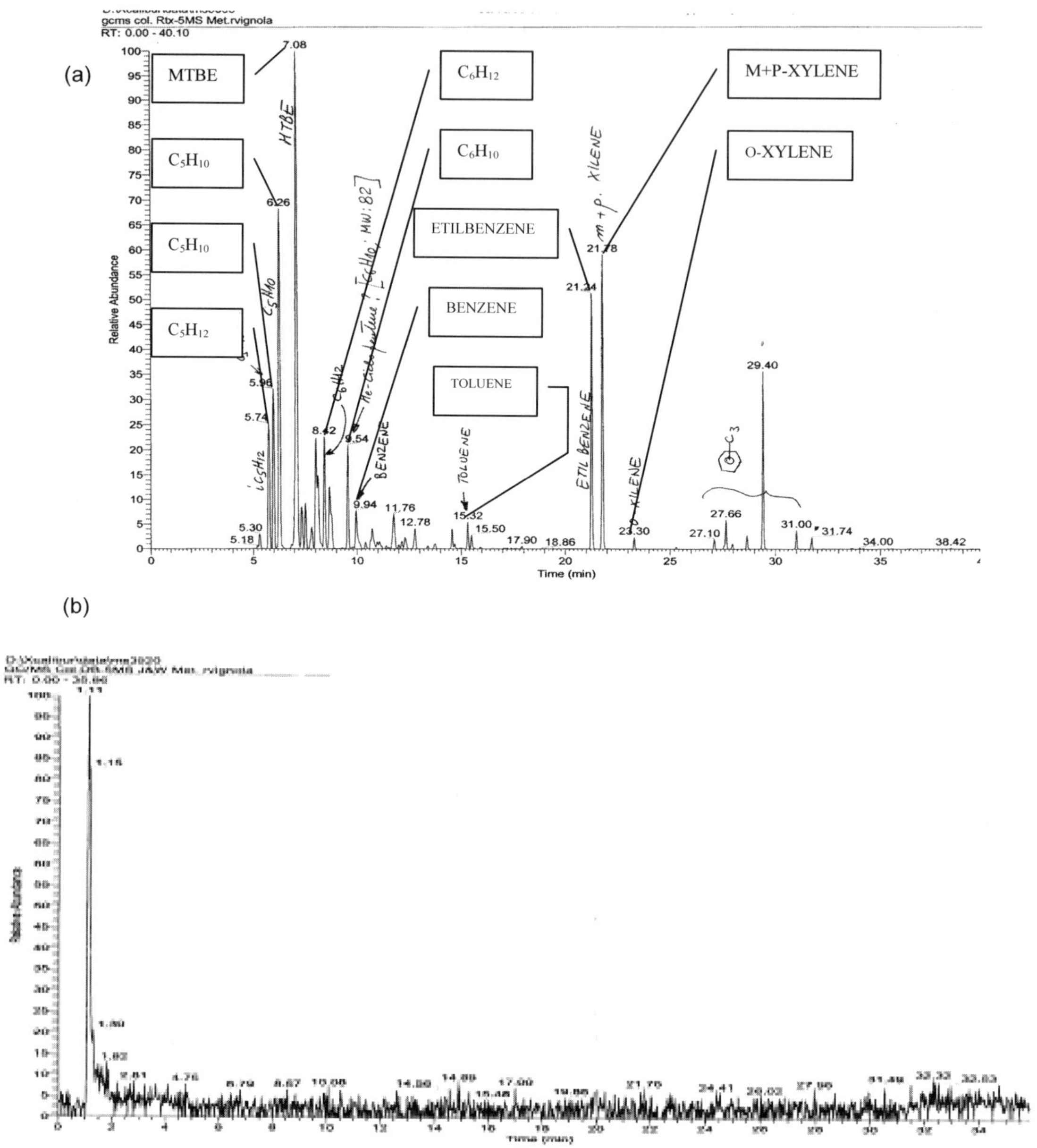

Fig. 2 Treatment of groundwater heavily contaminated by MtBE (17.9 mg l^{-1}) in presence of aromatic hydrocarbons: mainly benzene (2.8 mg l^{-1}), toluene (0.5 mg l^{-1}), ethylbenzene (1.5 mg l^{-1}), *o*-xylene (1.3 mg l^{-1}), *p*-xylene (2.9 mg l^{-1}). Salinity is 15 g l^{-1} as NaCl. (a) The inlet GC-MS profile; (b) the outlet profile. The amplification of outlet profile evidences that the contaminant concentrations are below analytical limits (about 1 μ g l^{-1}), and demonstrates the complete remediation.

In this case the outlet profile also shows that all the contaminants are immobilized in the zeolite adsorbent.

CONCLUSIONS

(a) HZ are suitable to adsorb all classes of organics and appear particularly advantageous in the treatment of compounds that are hard to remediate, such as 1,2-DCA, which is not easy to treat with either available (GAC, ZVI) nor advanced reactive media (Pd/H_2), and MtBE, whose concentration, even when higher than tens of mg l^{-1}, is easily reduced at levels <10 μg l^{-1}.
(b) Zeolite sorption capacity is not inhibited by the presence of high concentrations of ions commonly present in groundwater, even in coastal sites (GAC does not work effectively in brackish water).
(c) In all cases experimented, contaminant concentration at the outlet of zeolite adsorption is below detection limit.
(d) The expected performance of *in situ* PRB employing zeolites as active sorbent material show promising perspectives in terms of efficiency and cost, especially in situations with a complex contamination

Acknowledgements The authors would like to acknowledge Adriano Bernardi, Umberto Cova, Silvano Deli, Gino Della Penna and Vincenzo Mancini for their skilled contributions.

REFERENCES

EDAG, Engineering and Design Adsorption Guide (2001) P. N° DG 1110-1-2 (www.usace.army.mil/usace-docs/design-guides/dg1110-1-2/contents.htm).

Fajula, F. & Plee. D. (1994) *Stud. Surface. Sci. Cat.* **85**, 633–651.

Liang, S., Palencia, L. S., Yates, R. S., Davis, M. K., Bruno, J. & Wolfe, R. L. J. (1999) *Amer. Water Works Ass.* **91**, 104–114.

Scott, W. D. & Powers, S. E. (2000) *J. Environ Eng.* **April**, 354–360.

Schad, H., Haist-Gulde, B., Klein, R., Maier, D., Maier, M. & Schulze, B. (2000) Funnel-and-gate at the former manufactured gas plant site in Karlsruhe: sorption test results, hydraulic and technical design, construction. In: *Contaminated Soil 2000* (Proc. Seventh Int. FZK/TNO Conf. on Contaminated Soil, 18–22 September 2000, Leipzig, Germany) vol. 2, 951–959.

Watson, M. (1996) Treatment of alternatives for MTBE in groundwater. *Applied Research Department Technology, Transfer Note no. 11.*

Permeable Reactive Barriers (Proceedings of the International Symposium held at Belfast, Northern Ireland, March 2004). IAHS Publ. 298, 2005.

Suitability testing of selected wastes and natural substrates for permeable reactive barriers (PRBs)

JANTHAWAN WANTANAPHONG,
ELIZABETH HELEN BAILEY & SACHA JON MOONEY
Environmental Science, School of Biosciences, University of Nottingham, University Park, Nottingham NG7 2RD, UK
liz.bailey@nottingham.ac.uk

Abstract Performance of barrier materials is important for the long-term success of permeable reactive barriers (PRBs). A number of selected wastes and natural materials were tested in batch experiments for their ability to remove heavy metals from synthetic groundwater. Two potential materials (clinoptilolite and calcified seaweed) were further investigated in column and elution experiments. Results showed that both materials have high metal removal (immobilization) (>95%) efficiencies. However, pore clogging as a consequence of mineral precipitation was recorded in a calcified seaweed column (size fraction 0.1–1.0 mm) after 7 months. Clogging was not observed in clinoptilolite columns. Elution tests showed some subsequent release of Cu, Cd and Zn in clinoptilolite and calcified seaweed columns eluted with acidic solution. Desorption of Cd and Zn was observed in a clinoptilolite column eluted with alkaline solution. No significant release of metals was observed in any of the columns eluted with a neutral pH fluid.

Key words calcified seaweed; clinoptilolite; column experiments; elution experiments; heavy metals; permeable reactive barriers

INTRODUCTION

Effective and financially attractive materials are essential for commercially viable remediation technologies. Permeable reactive barriers (PRBs) provide an innovative approach for groundwater remediation. The vast majority of PRB systems use zero valent iron as a barrier material. However, potential limitations of iron are reduction in porosity, chemical reactivity and hydraulic conductivity over time due to clogging of pore spaces by corrosion and mineral precipitation (Liang *et al.*, 2003; Phillips *et al.*, 2003). Recently, the use of natural and waste materials for remediation has received attention as they provide a cheap alternative to synthetic products. It is evident from numerous studies that materials such as zeolite, calcified seaweed, chitin, clay, fly ash and dried seaweed have demonstrated outstanding capabilities for removal of heavy metals such as Pb, Cd, Cu and Zn from contaminated waters (Bailey *et al.* 1999; Babel & Kurniawan, 2003). This research considers selective waste products and natural materials for use as barriers.

EXPERIMENTAL DESIGN

Suitable barrier materials should exhibit high contaminant removal efficiencies and be able to treat a range of groundwater chemistries whilst maintaining an adequate porosity and permeability. They should also not subsequently release immobilized contaminants as a consequence of changes in local groundwater conditions. In order to examine these parameters, a series of batch and column experiments were performed on selected materials (Table 1). Clinoptilolite, calcified seaweed, cocoa shell, chitin flakes, clay mixtures and fly ash were preliminary screened in batch experiments. The experiments were conducted by mixing 0.8 g of material and 40 cm^3 synthetic groundwater (10 mg l^{-1} Pb, Cu, Zn, 1 mg l^{-1} Cd, pH 5) and removal efficiencies, pH and morphological change were determined. Suitable materials were then selected to investigate in column experiments over periods of 1–18 months at low (0.35 cm^3 min^{-1}) and high (2.08 cm^3 min^{-1}) flow rates with different synthetic groundwaters (Table 2) and with two fraction sizes (0.1–1.0 mm, 1.0–2.0 mm). The column was packed with sand at the bottom followed by reactive material with a final layer of sand. Samples taken from intermittent ports and the effluent exiting the column were periodically collected for metal analysis and pH determination. To mimic post treatment changes in groundwater condition, elution experiments were undertaken. Reacted columns were eluted with uncontaminated acidic (pH 2.5), alkaline (pH 9) or neutral (pH 7) groundwater at a flow rate of 2.08 cm^3 min^{-1} for 4 weeks. The metal-free fluid contained 50 mg l^{-1} Ca, 40 mg l^{-1} Na, 5 mg l^{-1} Mg, 3 mg l^{-1} K. At the conclusion of the tests, materials were collected for SEM-EDX analysis to: (a) observe morphology and identify mineral precipitates; and (b) quantify physical properties such as porosity, pore size and pore shape using image analysis (Wantanaphong *et al.*, 2005, this volume).

Table 1 Materials tested in batch experiments.

Materials	Source	Size fraction (mm)
Crushed clinoptilolite	Natural mineral	0.1–1.0
Calcified seaweed granules	Calcareous algae with coral like structure	1.0–2.0
Crushed cocoa shell	Waste product from chocolate industry	1.0–2.0
Clay mixtures	A mixture of 31% clay, 38% silt and 31% sand	0.1–1.0
Chitin flakes	Natural polymer (waste product from crab and shrimp shells)	0.3–1.0
Fly ash powder	Coal fly ash	<0.05

Table 2 Synthetic groundwater compositions used in this study. All concentrations in mg l^{-1}.

	Pb	Cu	Zn	Cd	Fe	Mn	Ca	Na	Mg	K	SO_4	Cl	HCO_3^-	CO_3^{2-}	NO_3^-
Batch Experiments															
pH 5.0:	10	10	10	1								10.84			26.59
Column Experiments															
Low Flow rate															
pH 3.5 :	10	10	10	1								10.84			26.59
High Flow rate:															
AMD (pH 2.5) :	10	10	10	1	0.5	85	500	800	550	100	3675	1237			
Neutral (pH 7) :	10	10	10	1			50	40	5	3	36	17	106	75	

RESULTS

Batch experiments

Clinoptilolite, calcified seaweed, chitin, clay, cocoa shell and fly ash were tested and proved effective for metal removal (Table 3). Chitin and fly ash performed exceptionally well, though chitin was expensive compared to other tested materials and hence less financially attractive for full-scale PRB application (assuming a similar metal removal performance to the other materials). Fly ash increased pH considerably, which favoured metal precipitation. Cocoa shell and clay mixtures resulted in low permeability systems that hindered flow, and hence effectiveness as *permeable* barriers. Clinoptilolite and calcified seaweed were selected for further investigation.

Clinoptilolite performed well under low flow-rate tests with complete removal of Cu, Cd, Pb and 98% removal of Zn during a 10 month experiment. No breakthrough was observed for any metal although the capacity of some column material to adsorb Cu and Zn was found to be exhausted when fluid samples were taken at intervals along the length of the column. In a column with a 10 cm reactive zone packed with the small particle size (0.1–1.0 mm), no pore clogging was observed. Results over 18 months indicated initial breakthrough of Zn, Cu (95 days) and then Cd (109 days) with complete exhaustion of column material after 137 days (Zn and Cu) and 312 days (Cd). Incomplete breakthrough was observed for Pb. Clinoptilolite adsorption capacity was 0.3–0.86 mg g^{-1} for Cd and 2.62–3.78 mg g^{-1} for Cu and Zn. As metal removal by clinoptilolite is predominantly by ion-exchange (Inglezakis *et al.*, 2002, 2003), elution experiments were performed to test the extent to which metals may be subsequently desorbed as a consequence of changing groundwater conditions. After reaction with metal-rich groundwater for four weeks, columns were subjected to flushing with an uncontaminated acidic, alkaline or neutral pH groundwater for a further four weeks. Results of the column experiments are presented in Table 4.

After the initial four week reaction with an acidic metal rich groundwater, complete breakthrough for Cu, Zn, Fe and Mn was observed. For Cd and Pb, 50% and 25% of the concentration in the synthetic groundwater was observed in the column effluent. No breakthrough of Cu, Cd and Pb was observed when a neutral pH groundwater was used. Breakthrough was observed for Zn with 70% of the input concentration observed in the column output after four weeks. Elution tests (Table 5) with an acidic eluent of the column initially treated with neutral pH contaminated water showed some desorption of Cu, Cd and Zn. After four weeks some 15%, 62%

Table 3 Batch test results after the reaction of materials with synthetic groundwater for 72 h.

Materials	Initial pH of solution	Final pH of solution	Removal efficiencies (%) Pb	Cd	Cu	Zn
Crushed clinoptilolite	5.9	8.0	98.6	99.8	98.4	99.7
Calcified seaweed	6.0	8.6	99.7	99.9	99.4	98.8
Crushed cocoa shell	5.4	6.5	87.5	90.5	77.9	70.2
Clay mixtures	5.2	7.3	98.7	92.6	97.8	95.0
Chitin flakes	5.6	7.6	99.9	99.8	98.8	100
Fly ash powder	5.0	9.5	100	100	100	100

Table 4 Metal sorption capacities of clinoptilolite and calcified seaweed. Values represent maximum sorption capacities except for low flow-rate columns where no or incomplete breakthrough was observed (values in italics). Where a range is quoted this represents capacities calculated at the point at which metals were first observed in column effluent and that when metal concentrations in the column effluent were the same as in the synthetic groundwater inflow.

	LOW FLOW RATE ($0.35\ cm^3\ min^{-1}$, Residence time ~15–19 h)				HIGH FLOW RATE ($2.08\ cm^3\ min^{-1}$, Residence time ~2.6 h)			
	Clinoptilolite	Clinoptilolite	Calcified seaweed	Calcified seaweed	Clinoptilolite	Clinoptilolite	Calcified seaweed	Calcified seaweed
Size fraction (mm)	0.1–1.0	1.0–2.0	0.1–1.0	1.0–2.0	1.0–2.0	1.0–2.0	1.0–2.0	1.0–2.0
Groundwater	Acidic pH= 3.5	Acidic pH= 3.5	Acidic pH= 3.5	Acidic pH= 3.5	AMD pH=2.5	Neutral pH=7	AMD pH=2.5	Neutral pH=7
Column height of sand/substrate/sand (cm)	20/10/20	5/40/5	20/10/20	5/40/5	20/10/20	20/10/20	20/10/20	20/10/20
Volume of substrate (cm^3)	152	608	152	608	152	152	152	152
Mass of substrate (g)	181	727	205	778	175	175	200	200
Duration of experiments	550 days breakthrough	301 days	221 days clogged	301 days	28 days breakthrough	28 days breakthrough	28 days breakthrough	28 days breakthrough
Sorption capacity ($mg\ g^{-1}$)								
Pb	> 15.19	*>2.07*	*>5.39*	*>1.93*	4.8	4.8	4.2	4.2
Cu	2.62–3.78	*>2.07*	*>5.39*	*>1.93*	4.8	4.8	4.2	4.2
Cd	0.30–0.86	*>0.21*	*>0.54*	*>0.19*	0.48	0.48	0.42	0.42
Zn	2.62–3.78	*>2.07*	*>5.39*	*>1.93*	4.8	4.8	4.2	4.2
Fe					0.24	–	0.21	–
Mn					40.8	–	35.7	–

* AMD = Acid mine water

Table 5 Release of metals during elution of clinoptilolite and calcified seaweed columns.

Metal	Total metal sorbed (mg)	% Metal release (dissolution/desorption)					
		Neutral GW eluted with acidic fluid		Neutral GW eluted with alkaline fluid		Acidic GW eluted with neutral fluid	
		1 week	4 weeks	1 week	4 weeks	1 week	4 weeks
Clinoptilolite							
Cu	840	5.30	15.00	0.04	0.12	0.04	0.22
Cd	84	20.75	62.43	4.41	20.26	0.18	1.65
Pb	840	0.33	2.21	0.31	0.99	0.31	2.64
Zn	840	13.86	54.89	4.32	6.65	0.19	0.78
Fe	42	No Fe				2.25	5.13
Mn	7140	No Mn				0.06	0.11
Calcified seaweed							
Cu	840	0.37	9.97	0.01	0.05	0.33	3.19
Cd	84	0.92	13.28	0.06	2.10	0.21	1.81
Pb	840	0.32	2.05	0.10	0.47	0.21	1.82
Zn	840	30.12	94.98	0.01	0.53	0.27	2.16
Fe	42	No Fe				1.13	3.80
Mn	7140	No Mn				0.09	0.79

and 55% of initially adsorbed Cu, Cd and Zn (respectively) had been desorbed. Some desorption of Cd and Zn was also observed in an identically treated column eluted with an alkaline solution. No significant desorption of metals was observed in columns eluted with a neutral pH fluid.

Column experiments

Calcified seaweed performance under low flow-rate conditions (over 10 months) was excellent with high removal efficiency observed for Cu, Cd and Zn (nearly 100%) and Pb (95%). Concentrations of metals measured along the column showed that after nine months, 50% saturation of Cu, Cd and Zn was observed at a distance of 5 cm (76 cm^3) indicating that the remaining 35 cm (532 cm^3) of column was still effective. Pore clogging by metal carbonate precipitates (Fig. 1) was observed in a column packed with the small particle size (0.1–1.0 mm) material after 7 months. After four weeks, reaction with acidic groundwater showed (at the high flow-rate) 100% breakthrough for Zn and Mn. No breakthrough was observed in a column treated with neutral pH groundwater. Elution results (Table 5) of this column with an acidic fluid showed significant desorption of Zn over the first week (30%), with 95% of the total adsorbed Zn desorbed within four weeks. Desorption of Cu and Cd was not high in the first week but gradually increased to 10–13% after four weeks. No desorption of metals was observed in the columns eluted by neutral and alkaline pH solutions.

DISCUSSION

The long-term success of PRBs depends upon the physical and chemical properties of barrier materials. Potential barrier materials should maintain a reactive surface and

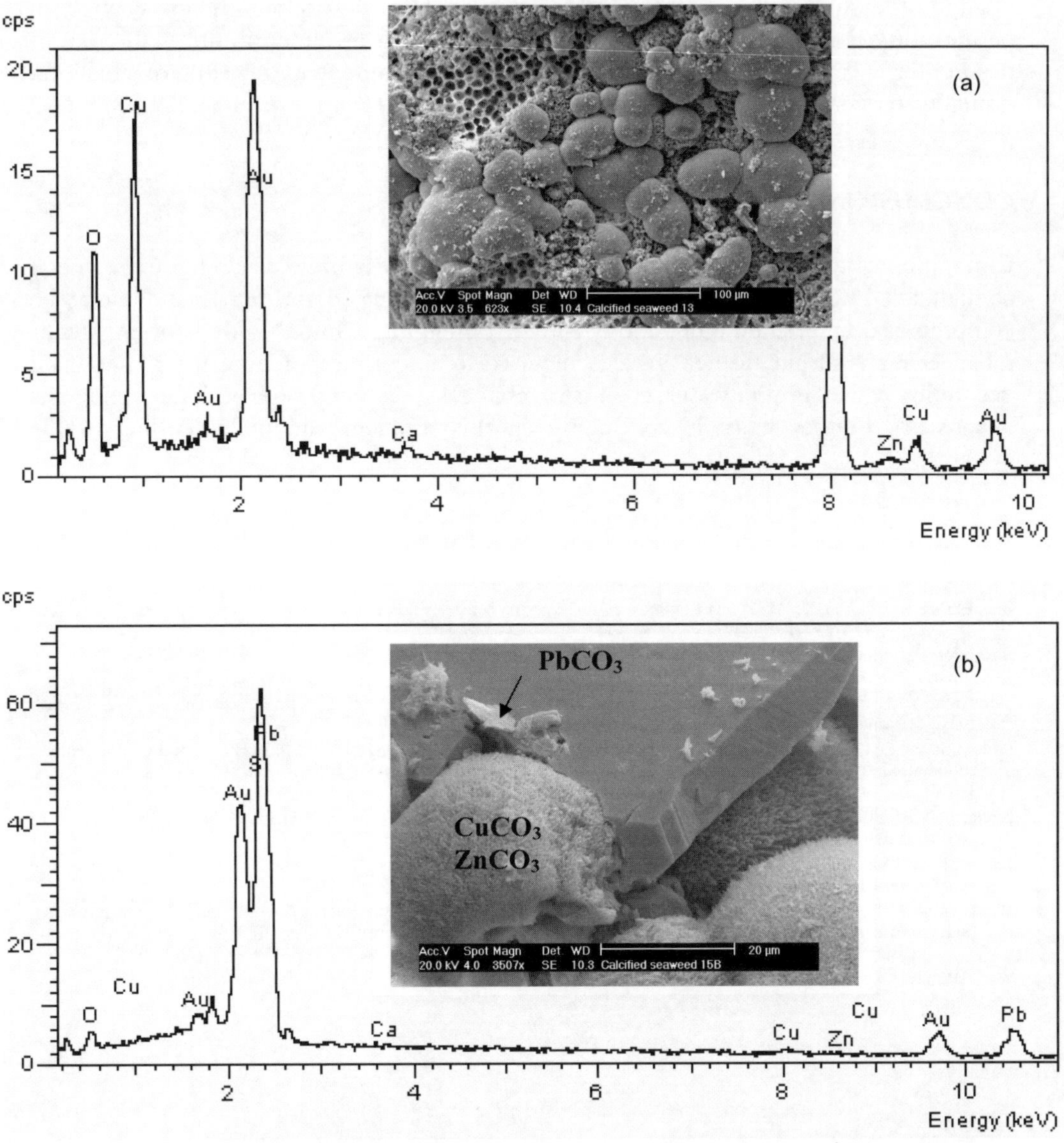

Fig. 1 SEM images and EDS spectra of calcified seaweed column material after treatment at low flow-rate with an acidic metal contaminated groundwater showing metal carbonate precipitates: (a) after 10 months, (b) after 221 days, clogged column.

appropriate permeability for a long time. Metal removal by clinoptilolite and calcified seaweed were excellent at low flow-rates and at high flow-rate conditions when residence times of fluid in the columns was shorter. Permeability reduction in calcified seaweed columns using the small size fraction (0.1–1 mm) calcified seaweed led to a short-life time of the system due to precipitation (Fig. 1). Mineral precipitates included Pb, Cu and Zn carbonates. Ion-exchange is the major metal removal mechanism in clinoptilolite and it can react for longer periods with no significant change in permeability or pore clogging. However, the limitation of clinoptilolite is the risk of desorption of metals back to the surrounding environment as a result of changing groundwater conditions.

CONCLUSIONS

Clinoptilolite and calcified seaweed can be successfully used to treat metal contaminated waters. However, the use of calcified seaweed as a barrier alone may be compromised by precipitation, and eventually, clogging. Clinoptilolite is promising as a barrier material but further work is required to assess the potential for contaminant desorption when groundwater chemistry changes (e.g. after the pollutant plume is exhausted). Further work is continuing on this problem and on modelling of the systems.

REFERENCES

Babel, S. & Kurniawan, T. A. (2003) Low-cost adsorbents for heavy metals uptake from contaminated water: a review. *J. Hazard. Mater.* **97**, 219–243.

Bailey, S. E., Oline, T. J., Bricka, R. M. & Adrain, D. D. (1999) A review of potentially low cost-sorbents for heavy metals. *Water. Res.* **33**, 2469–2479.

Inglezakis, V. J., Loizidou, M. D. & Grigoropoulou, H. P. (2002) Equilibrium and kinetic ion exchange studies of Pb^{2+}, Cr^{3+}, Fe^{3+} and Cu^{2+} on natural clinoptilolite. *Water Res.* **36**(11), 2784–2792.

Inglezakis, V. J., Loizidou, M. D. & Grigoropoulou, H. P. (2003) Ion exchange of Pb^{2+}, Cu^{2+}, Fe^{3+} and Cr^{3+} on natural clinoptilolite: selectivity determination and influence of acidity on metal uptake. *J. Colloid. Interface Sci.* **261**(1), 49–54.

Liang, L. Y., Sullivan, A. B., West, O. R., Moline, G. R. & Kamolpornwijit, W. (2003) Predicting the precipitation of mineral phases in permeable reactive barriers. *Environ. Engng Sci.* **20**, 635–653.

Philips, D. H., Watson, D. B., Roh, Y. & Gu, B. (2003) Mineralogical characteristics and transformations during long-term operation of a zerovalent iron reactive barrier. *J. Environ. Qual.* **32**(6), 2033–2045.

Wantanaphong, J., Bailey, E. H. & Mooney, S. J. (2005) Quantification of the effects of pore clogging in potential PRB substrates using image analysis. In: *Permeable Reactive Barriers* (ed. by G. Boshoff & B. D. Bone) (Proc. Int. Symp., Belfast, Northern Ireland, March 2004), 154–160. IAHS Publ. 298, IAHS Press, Wallingford, UK (this volume).

Remediation of acidic colliery spoil leachate in a hybrid passive treatment system comprising a permeable reactive barrier, ponds and a reedbed (Shilbottle, Northumberland, UK)

PAUL L. YOUNGER & MAHMOUD MOUSTAFA
School of Civil Engineering and Geosciences, Drummond Building, University of Newcastle, Newcastle upon Tyne NE1 7RU, UK
paul.younger@ncl.ac.uk

Abstract Highly acidic spoil leachates at the former Shilbottle Grange Colliery site (Northumberland, UK) have been constrained to flow through a permeable reactive barrier (PRB) containing a mixture of limestone gravel, horse manure and composted green waste. Leachates leave the PRB through a seepage face, draining into ponds and a reedbed. Iron concentrations decrease across the system from >800 mg l^{-1} to ~10 mg l^{-1}. Iron loading reduction is about 98%. While pH rises from 4 to 6 in the PRB, subsequent precipitation of metals as hydroxides and hydroxysulphates leads to the apparent paradox of simultaneous decreases in both total acidity and pH. The PRB removes about 30% of the sulphate (by reductive precipitation of metal sulphides), while the lagoons and reed bed together remove another 20% (as hydroxy-sulphate precipitates). The system as a whole reduces the sulphate content of the raw water by 50%.

Key words acidic mine drainage; geochemistry; groundwater; mining; permeable reactive barrier; pollution; remediation; treatment; spoil; UK; wetlands

INTRODUCTION

It is now clear that the most effective passive treatment pathways for acidic mine waters are those based upon anoxic geochemical processes, including carbonate dissolution in the absence of Fe^{3+}, and/or bacterial sulphate reduction (BSR) (Younger *et al.*, 2002). Both of these processes generate bicarbonate alkalinity and raise pH; BSR also leads to immobilization of ecotoxic metals by the precipitation of metallic sulphides. It is easiest to achieve and maintain conditions conducive to these processes by deploying subsurface flow through porous media (Younger *et al.*, 2002). Some of the challenges involved in engineering systems based on these principles include predicting and prescribing the permeability of manure-based composts; matching sufficiently high permeability with a high enough organic matter content to ensure microbial activity; and geometrical considerations in systems design. One example of how these challenges are overcome in practice is provided by the Shilbottle passive treatment system in Northumberland (UK); a unique installation incorporating a large permeable reactive barrier (PRB) and a series of open-air ponds and a reedbed (see Fig. 1). The principles of using sedimentation ponds and reedbeds to passively treat polluted mine drainage have been widely discussed in the recent literature (see Younger *et al.*, 2002, for a review, and PIRAMID Consortium 2003 for practical

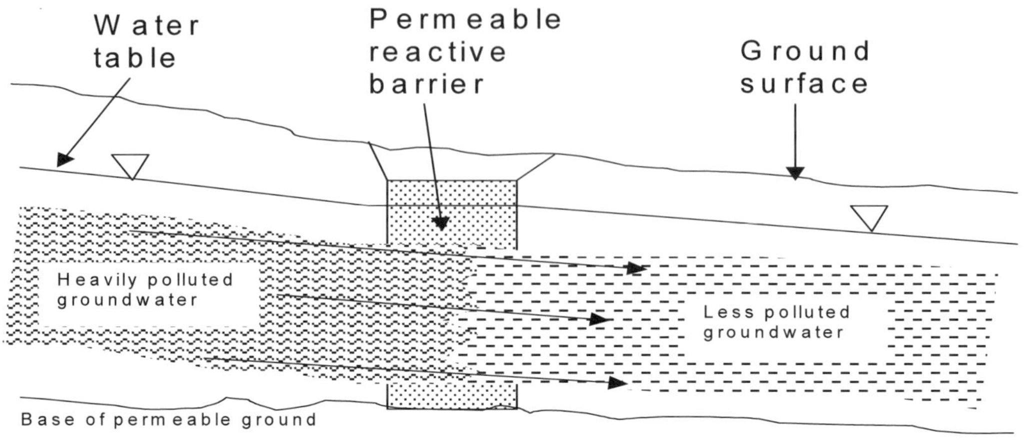

Fig. 1 Sketch cross-sections of a permeable reactive barrier (PRB) (after Younger, 2000).

guidance on wetland treatment engineering). In this paper we focus on the PRB element of the overall treatment system at Shilbottle, emphasizing the contribution it makes to overall treatment of the highly acidic leachates at that site.

PRB TREATMENT OF ACIDIC MINE WATERS

Permeable reactive barriers (PRBs) are increasingly being employed to treat a wide range of polluted groundwaters (e.g. Gavaskar *et al.*, 1998; Beck *et al.*, 2001), including acidic mine waters (e.g. Benner *et al.*, 1997; Younger *et al.*, 2002). For the latter application the current favoured approach is to use a reactive medium containing compost or some other carbon sources, to promote bacterial sulphate reduction, which tends to lead to neutralization of the water and removal of metals from solution by precipitating them as sulphides. As in other passive technologies employing this type of approach (e.g. compost wetlands and so-called "RAPS" bioreactors; Younger *et al.* 2002), the substrates used in mine water treatment PRBs often also include limestone to provide inorganic neutralisation capacity.

THE SHILBOTTLE PASSIVE TREATMENT SYSTEM

Pyritic shale brought to the surface during mining at Shilbottle Grange Colliery was deposited in a large heap, which emits extremely acidic leachates. These have been characterized by repeated sampling of a field of monitoring wells installed along the southernmost flank of the spoil heap (Table 1). The extreme contaminant concentrations in these groundwaters qualify the Shilbottle spoil leachates as some of the most contaminated yet documented in England. Discharge of this leachate to a nearby stream (the Tyelaw Burn) has long resulted in spectacular pollution, as various metals are precipitated in the form of hydroxides, such as bright orange iron ochre,

Table 1 Hydrochemical parameters (total concentrations) for six monitoring boreholes (BH6 to BH1; numbering decreases west to east) and three surface seepages of groundwater (GW 9–11) at the Shilbottle Spoil Heap site, Northumberland, UK (Younger *et al.*, 2003).

	BH 6	BH 5	BH 4	BH 3	BH 2	BH 1	GW 11	GW 10	GW 9
pH	4.47	3.64	4.00	4.08	4.17	6.32	3.29	3.55	4.17
Acidity (mg l^{-1} CaCO3)	742	2048	2557	2908	6342	78	1360	2534	3322
Fe (mg l^{-1})	92	200	405	599	1136	26	278	452	688
Mn (mg l^{-1})	68	108	180	205	299	22	165	181	238
Al (mg l^{-1})	86	267	270	263	678	1.2	97	249	298
SO_4^{2-} (mg l^{-1})	8701	8162	8230	10167	15318	3745	6334	9288	11176

white aluminium hydroxysulphate foam and (locally) black manganese "wad" with an iridescent sheen. These precipitates are not only unsightly; they also smother the benthos and thus prevent algal photosynthesis. Furthermore, the precipitation of these hydroxides releases protons, depressing the pH even further and rendering the Burn extremely inhospitable to normal aquatic fauna and flora. Thus the Shilbottle leachate has caused severe ecological degradation for several decades, and has affected water quality to such an extent that the impacts remain significant even after major dilution downstream of the confluence of the Tyelaw Burn and the River Coquet.

Initial surface reclamation successfully established a vegetative cover on the spoil. However, attempts to improve the quality of visible surface seepages of leachate using reedbeds were hindered by bypass flows of >60% of the leachate to the stream and by a lack of alkalinity-generating processes within the spoil (Younger *et al.*, 2003). Given the low hydraulic heads across the site, technical options for improving passive treatment were limited. An innovative design developed by Northumberland County Council and the University of Newcastle (Amos & Younger, 2003) led to installation of a permeable reactive barrier, on the downstream face of which the groundwater surfaces and flows into a series of three open ponds. Further removal of oxidizable metals then occurs, prior to final "polishing" of the effluent in the pre-existing reed-bed (Fig. 2). The Shilbottle PRB is by far the largest of its type in the world, being 180 m long (N–S), 3 m deep by 2 m wide (Fig. 2). It was constructed by a cut-and-fill method, in which a trench was first excavated the full length of the barrier before tipping of pre-mixed reactive media. In line with the recommendations from extensive laboratory-testing of possible reactive media (Amos & Younger, 2003), the Shilbottle PRB substrate consists of 25% composted horse manure and straw, 25% green waste compost, and 50% limestone gravel (mostly 25 mm nominal single-sized, though there is one section of the barrier with 10 mm diameter limestone gravel). In one section of the PRB, the limestone clasts were substituted by blast furnace slag, to allow comparison of this recycled source of potential alkalinity with natural limestone, which has to be quarried. These organic and inorganic components of the substrate were thoroughly mixed using a large, mobile agricultural mixer. The mix was then loose-tipped into the trench to the full depth of excavation. This created a substantial unsaturated zone of reactive media above the water table in the spoil, which both provides further reactive capacity for some future period when heads in the PRB rise in response to clogging of pores, and also provides an O_2 consumption blanket in the meantime. Multi-level piezometers were installed in the media in two positions during

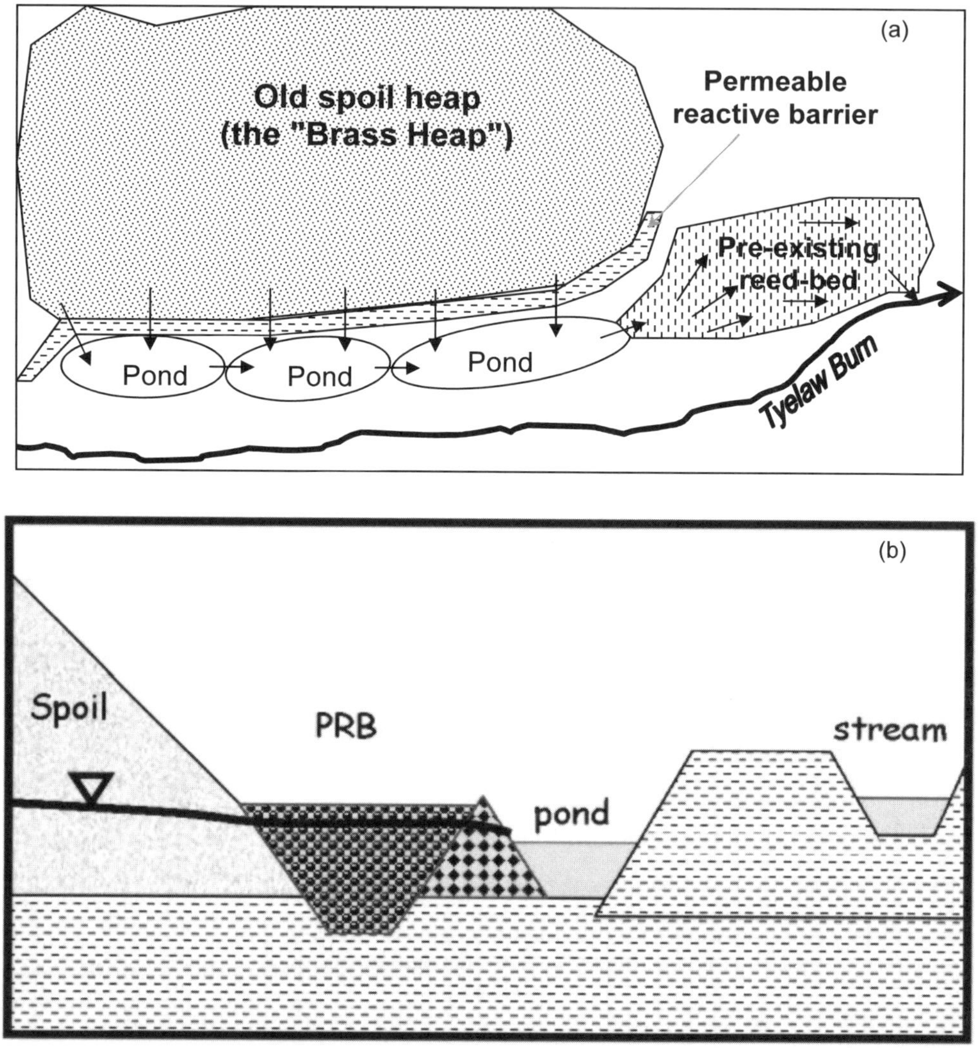

Fig. 2 (a) Sketch map and (b) schematic cross-section of the hybrid PRB-ponds-wetland system at Shilbottle, Northumberland, UK. Black arrows show selected approximate water flow directions.

emplacement and a further set of sampling points have now been established, complementing the previously existing boreholes up gradient of the PRB (previously sampled to yield the data in Table 1) that continue to provide information on the native groundwater in the spoil. The downstream face of the PRB was lined with brick rubble, to provide a permeable exit filter to the oxidation ponds. In some places, an artificial liner was draped along the downstream face to divert some of the more voluminous feeders of water through larger volumes of the permeable reactive media than they would have been likely to enter had they been permitted to take the shortest flowpath across the barrier.

PERFORMANCE OF THE SHILBOTTLE PASSIVE TREATMENT SYSTEM

The Shilbottle passive treatment system has now been operational for a little more than one year. Figure 3 summarizes treatment performance in relation to Fe and pH. The removal of Fe is extremely impressive, going from >800 mg l^{-1} in the raw water down to only 10 mg l^{-1} at the final outfall to a nearby stream (the Tyelaw Burn). Compared with the situation before installation of the PRB and ponds, this removal rate equates to a net reduction in release of Fe to the Tyelaw Burn of 98%. The pH rises from 4 to 6 in the PRB, but given the extreme acidity of this water, insufficient alkalinity can be added to completely neutralise the mineral acidity within the PRB itself. However, this temporary increase in pH speeds up the subsequent precipitation of metals as hydroxides and hydroxysulphates (which happens thousands of times faster at pH 6 than at pH 4). As this occurs, the alkalinity provided by the PRB is consumed and the pH drops once more. This results in the apparent paradox of simultaneous decrease in

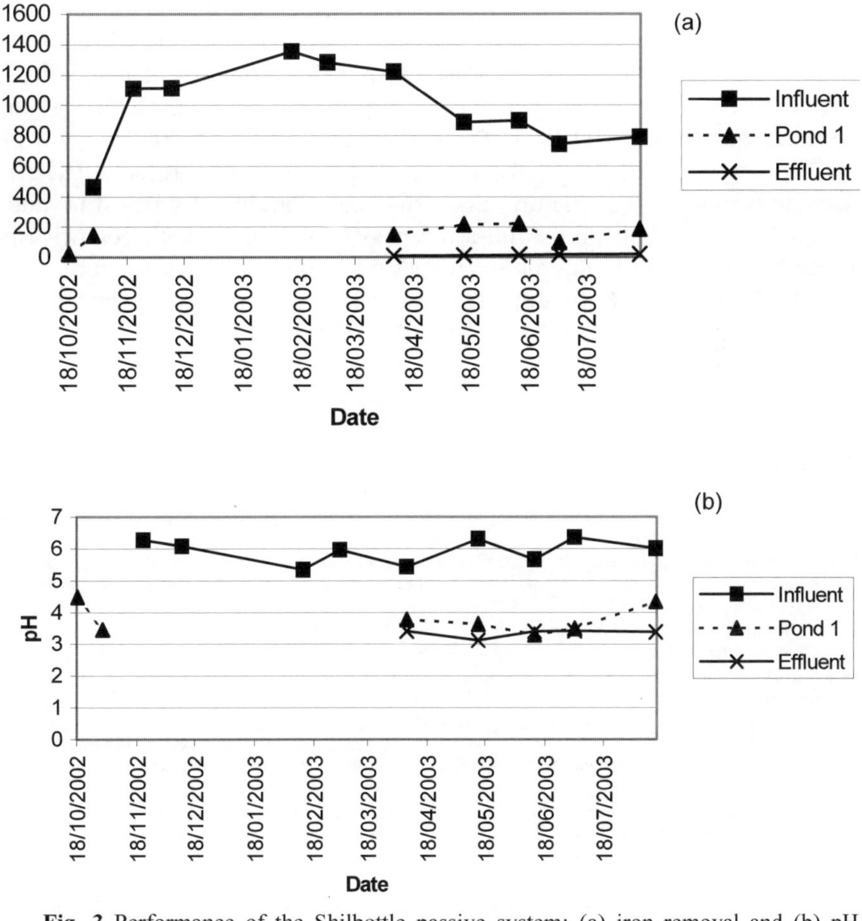

Fig. 3 Performance of the Shilbottle passive system: (a) iron removal and (b) pH removal, between the groundwaters entering the PRB ("influent"), water in the oxidation ponds ("Pond 1") and the final effluent from the reedbed ("effluent").

both total acidity and pH: mineral acidity is comprehensively removed from the water as it flows through the treatment system, but this results in a lowering of pH due to ferric iron hydrolysis. The low pH at the end of the wetland is now only weakly buffered by the remaining 10 mg l^{-1} Fe. Future plans are to install semi-passive soda ash dosing within the final wetland to raise pH to 7 and speed up the removal of the remaining Fe. In terms of sulphate removal, the system is performing better than analogous compost wetlands elsewhere in the UK and the USA (cf. Younger *et al.*, 2002). The PRB removes about 30% of the sulphate by reductive precipitation of metal sulphides. The lagoons and reedbed together remove another 20% as hydroxy-sulphate precipitates. The system as a whole therefore reduces the total sulphate content of the raw water by 50%. The net result of the treatment afforded by this system is that the Tyelaw Burn is now free of iron staining, and has a circum-neutral pH throughout its course.

Acknowledgements The authors gratefully acknowledge the collaboration of Aln Elliot of Northumberland County Council in getting the Shilbottle Permeable Reactive Barrier constructed, and the following agencies for funding the work: Northumberland County Council, One North East, European Regional Development Funding, Northumberland Environmental Partnership, SITA, Entrust, CL:AIRE, the European Commission's PIRAMID R&D Project (EVK1-CT-1999-000021; www.piramid.org), and the new DTI/BBSRC/NERC/EPSRC Biorem 4 LINK project "ASURE", which is also supported by Rio Tinto, Scottish Coal, Parkhill Estates Ltd and IMC Ltd. Shilbottle is now one of six full-scale passive systems which comprise the CoSTaR research facility, an international access research infrastructure sponsored by CL:AIRE, with access funding for researchers throughout Europe provided by the European Commission's 6th Framework RTD programme (contract no. RITA-CT-2003-506069).

REFERENCES

Amos, P. W. & Younger, P. L. (2003) Substrate characterization for a subsurface reactive barrier to treat colliery spoil leachate. *Water Res.* **37**, 108–120.

Beck, P., Harries, N. & Sweeney R. (2001) Design, installation and performance assessment of a zero valent iron permeable reactive barrier in Monkstown, Northern Ireland. *CL:AIRE Technology Demonstration Report TDP1.* (ISBN 0-9541673-0-9). CL:AIRE, London, UK.

Benner, S. G., Blowes, D. W. & Ptaceck, C. J. (1997) A full-scale porous reactive wall for prevention of acid mine drainage. *Ground Water Monitoring and Remediation* **17**(4), 99–107.

Gavaskar, A. R., Gupta, N., Sass, B. M., Janosy, R. J. & O'Sullivan, D. (1998) *Permeable Barriers for Groundwater Remediation: Design Construction and Monitoring.* Battelle Press, Columbus, Ohio, USA.

PIRAMID Consortium (2003) *Engineering guidelines for the passive remediation of acidic and/or metalliferous mine drainage and similar wastewaters.* European Commission fifth Framework RTD Project no. EVK1-CT-1999-000021 "Passive *in-situ* remediation of acidic mine/industrial drainage" (PIRAMID). University of Newcastle Upon Tyne, Newcastle Upon Tyne, UK.

Younger, P. L. (2000) Holistic remedial strategies for short- and long-term water pollution from abandoned mines. *Trans. Institution of Mining and Metallurgy (Section A: Mining Technology)* **109**, A210–A218.

Younger, P. L., Banwart, S. A. & Hedin R.S. (2002) *Mine Water: Hydrology, Pollution, Remediation.* Kluwer Academic Publishers, Dordrecht, The Netherlands.

Younger, P. L., Jayaweera, A., Elliot, A., Wood, R., Amos, P. Daugherty, A. J., Martin, A., Bowden, L., Aplin, A. C. & Johnson, D. B. (2003) Passive treatment of acidic mine waters in subsurface-flow systems: exploring RAPS and permeable reactive barriers. *Land Contamination and Reclamation* **11**(2), 127–135.

3 Monitoring of PRBs

Performance of multiple gate funnel-and-gate systems in three-dimensional, anisotropic, heterogeneous aquifers

CLAUDIUS BÜRGER & MICHAEL FINKEL

Center for Applied Geoscience, University of Tuebingen, Sigwartstr.10, D-72076 Tuebingen, Germany

claudius.buerger@uni-tuebingen.de

Abstract A synthetic modelling study was carried out to investigate performance differences between single and multi-gate funnel-and-gate systems (FGS) in 10 different anisotropic, heterogeneous aquifers. A FGS design model was used to obtain nine different FGS designs which are hydraulically equivalent based on a homogeneous aquifer. By means of a geostatistical description of anisotropy and heterogeneity of hydraulic conductivity, Monte-Carlo analyses were performed to obtain capture zone reliability curves and total gate flux distributions for the individual designs and aquifer scenarios. In combination with a simple cost model, reliability–cost plots were generated to evaluate design performance for both objectives. The results showed that depending on the actual distance upgradient of the FGS, multi-gate designs may capture a certain plume width more reliably than single gate designs. In terms of cost, the variance of the hydraulic conductivity field seems to have decisive influence concerning the cost-efficiency of multi-gate designs.

Key words aquifer heterogeneity; anisotropy; costs; design-reliability; funnel-and-gate system

INTRODUCTION

Permeable reactive barriers (PRB) are no longer considered innovative (e.g. Richards, 2002), but a proven conventional method for *in situ* groundwater remediation. Nevertheless, implementation and monitoring details still remain subject to scientific investigation. The advantageous idea of an *in situ* reactor that degrades or immobilizes harmful contaminants without energy consumption (as these are carried along by the natural groundwater flow) poses a disadvantageous dependency of the technology itself on the local groundwater flow system. Apart from seasonal deviations of ambient hydraulic gradient, the local flow and transport patterns are governed by the local geology or hydro-facies pattern of the aquifer. Unfortunately, these generally asymmetric hydro-facies patterns and related parameter values, e.g. hydraulic conductivity, cannot be assessed exhaustively throughout an area of interest due to obvious economic and technical constraints. Therefore, planners must base their decisions on an inevitably uncertain description of the flow field when designing a PRB. As subsequent corrections to a PRB already installed in the ground are both technically laborious and expensive, the reliability of performance predictions needs to be significantly higher than for an alternative active measure like pump-and-treat where upgrading is comparatively easily achieved.

An evaluation of PRB performance reliability and cost-efficiency can be carried out by use of geostatistical methods in combination with numerical groundwater models. Typically, the impact of a particular PRB design is studied for different geostatistically generated input parameter fields, e.g. hydraulic conductivity field realisations, and the model predictions are likewise evaluated according to their ensemble characteristics (Monte-Carlo Analysis). In this context the heterogeneity of the hydraulic conductivity is often modelled as a stationary multi-Gaussian random field with a spatial correlation structure, e.g. Koltermann & Gorelick (1996). Focusing on aquifers consisting of fluvial deposits, a directional hydro-facies pattern asymmetry can be expected according to the river sinuosity and its overall flow direction (e.g. the valley axis). This directional hydro-facies pattern asymmetry and likewise anisotropy of hydraulic conductivity should have an impact on the size of the capture zone of a PRB, as it renders certain flow directions more favourable than others. This can be modelled by relative scaling of the principal axes of the spatial correlation structure of the multi-Gaussian random field (e.g. Deutsch & Journel, 1992).

In this paper we investigate the impact of the anisotropy of the hydraulic conductivity on performance reliability of funnel-and-gate systems (FGS) with a zero valent iron degradation reactor. Single, 2-gate and 4-gate FGS with three different aspect ratios (total funnel width/total gate width, see Fig. 1 for details) are compared which show an "equivalent" performance in a homogeneous aquifer. The cost-efficiency of any particular design is finally evaluated by relating its reliability with respect to plume capture to its installation costs.

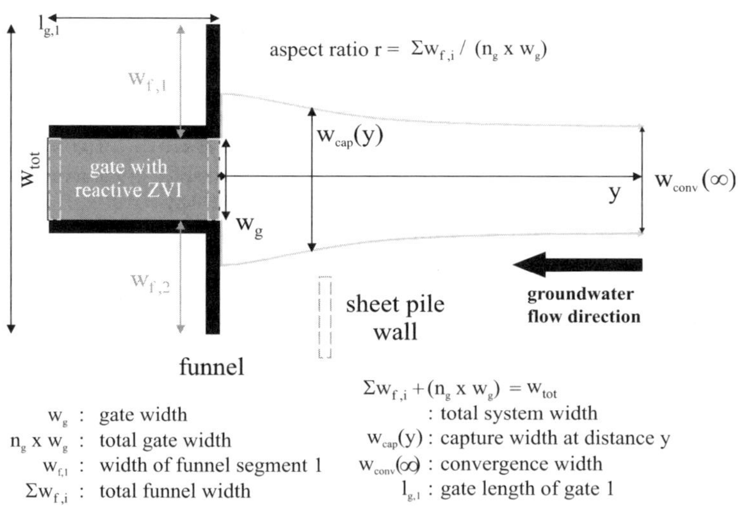

Fig. 1 Terms and definitions used for the funnel-and-gate system (FGS) design model.

METHODS

The determination of the sensitivities of FGS capture, reliability and cost was based on the following models and their interplay: a finite-difference flow model, a FGS-design model, a geostatistical aquifer model, and a simple cost model.

Flow model and capture zone evaluation

The US Geological Survey groundwater flow model MODFLOW-2000 (Harbaugh *et al.*, 2000) was used in combination with ZONEBUDGET 2.1 (Harbaugh, 1990) and MODPATH 4.3 (Pollock, 1994) to determine FGS' capture zones by particle tracking, assuming conservative transport and hydrodynamic dispersion to be negligible. The model domain (Fig. 2) consists of a grid defined by 372 rows, 372 columns and seven layers with varying horizontal increments (d_x, d_y) and 0.2 m increments for the vertical direction (d_z). Confined, steady-state groundwater flow was considered. The constant flow boundary guarantees for the same total water budget (specific discharge $q = 3.0 \times 10^{-7}$ m s^{-1}) independent of realisation or FGS design. In order to establish uniform inflow across the northern boundary, the northernmost nine grid rows were set to the geometric mean of the underlying geostatistical model. A constant head boundary (h = 20 m) was set at the southern boundary to ensure confined aquifer conditions.

The ensemble capture zone was determined by particle forward tracking for a total number of 250 equally-probable hydraulic conductivity realisations per aquifer scenario and FGS design. Particles were placed within an area upgradient of the FGS as illustrated in Fig. 1. Particles travelling through gate cells during the forward tracking process were stopped and the cell of their starting position marked as part of the capture zone. Capture zone evaluation was performed for any individual realization. The capture zone width, w_{cap} was quantified as twice the minimum distance of the capture zone boundary to the symmetry plane of the FGS system for each cell increment upstream of the FGS. This results in discrete values, $w_{cap}(y)$ for each model layer. The ensemble results of capture zone evaluation were then interpreted in terms of the reliability to achieve a certain capture zone width in a particular layer in a distance y from the FGS.

Investigated FGS designs

Only straight and symmetric FGS were considered. If more than one gate was present, all the gates were set at the same width. Funnels and gates fully penetrated the aquifer

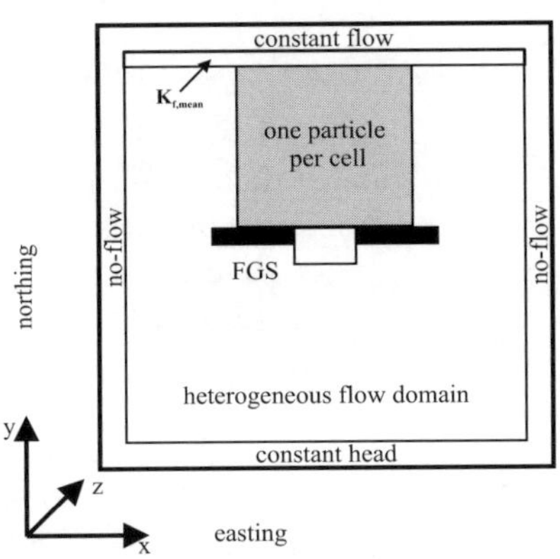

Fig. 2 Sketch of the numerical flow model set-up.

and are keyed to an underlying aquitard. Based on that, an FGS design can be described by its total width, w_{tot}, its aspect ratio r, the number of gates n_g, their relative positions, and their individual gate lengths $l_{g,i}$ (see Fig. 1 for illustration and explanation). Funnels were modelled as horizontal flow barriers of a hydraulic conductivity of 1.0×10^{-9} m s^{-1}. For the iron filling of the gates, a hydraulic conductivity of 0.01 m s^{-1} was assumed.

We compared FGS designs that would have an "equivalent hydraulic performance" in the case of a homogeneous aquifer. The two FGS designs were considered to be hydraulically equivalent if both had the same convergence width w_{conv} (see Fig. 1 and refer to Teutsch *et al.* (1997) and Bürger *et al.* (2003) for more information). The way the model domain was set up implies the same total gate flux for all the FGS designs under comparison.

As any convergence width can be obtained by an infinite number of FGS designs, the design parameter space was restricted by setting the number of gates n_g to 1, 2, and 4 and focused on three reasonably different aspect ratios r (2.1, 5.2, 14.5). For each of the nine FGS designs, the gate positions with the highest relative convergence width, $w_{conv,rel}$ (convergence width w_{conv}/total system width w_{tot}) were chosen. With n_g, r, and gate positions fixed the total system width w_{tot} was easily determined for any targeted convergence width. In this study w_{conv} was set to 50 m. Figure 3 shows a scaled sketch of the FGS designs under investigation (see Table 1 for the corresponding w_{tot} for w_{conv} = 50 m). Individual designs were referenced by the following abbreviations, e.g. 1G021, where 1 represents an n_g of 1 and 021 represents an r of 2.1.

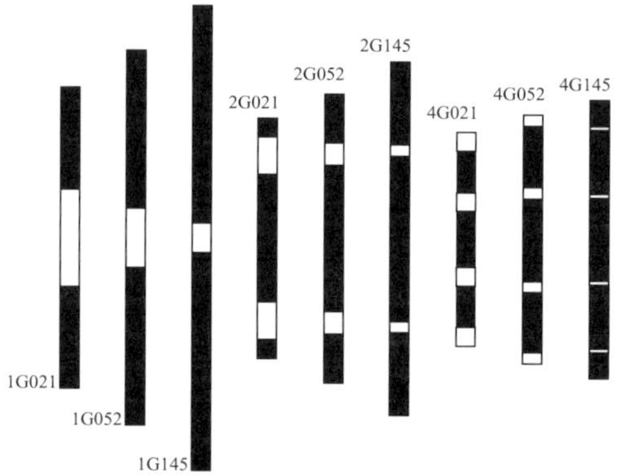

Fig. 3 Scaled sketch of the nine funnel-and-gate systems under comparison.

Table 1 Total system widths w_{tot} and identifiers for the FGS designs under investigation.

FGS-Designs	w_{tot} (m)
1G021	79.97
1G052	100.14
1G145	125.00
2G021	64.67
2G052	77.39
2G145	93.87
4G021	55.90
4G052	63.77
4G145	74.08

Table 2 Geostatistical model parameters used to model the 10 scenarios of aquifer heterogeneity.

Aquifer scenario	$K_{f,mean}$ (m s^{-1})	σ^2 (–)	λ_x (m)	λ_y (m)	λ_z (m)	φ (°)
iso	0.0001	1.0	14.14	14.14	1.0	0
long	0.0001	1.0	10	20	1.0	0
angle	0.0001	1.0	10	20	1.0	45
short	0.0001	1.0	10	20	1.0	90
v45_iso	0.0001	4.5	14.14	14.14	1.0	0
v45_long	0.0001	4.5	10	20	1.0	0
v45_short	0.0001	4.5	10	20	1.0	90
d_iso	0.0001	1.0	28.28	28.28	1.0	0
d_long	0.0001	1.0	20	40	1.0	0
d_short	0.0001	1.0	20	40	1.0	90

Geostatistical model

A total number of ten different heterogeneous, anisotropic aquifer scenarios was modelled. The corresponding geostatistical model parameters are presented in Table 2. The scenarios were not based on specific natural systems. Rather, their parameter range was oriented on reported results from field analyses in the scientific literature, e.g. Dagan, 1989; Elder *et al.*, 2002. The relative importance of aquifer heterogeneity depends on the ratio between the absolute lengths of the anisotropic structures and the size of the problem under consideration (here: the convergence width for which the FGS is designed). If the required capture zone width, and respectively, the total FGS width are much larger than the scale of the heterogeneities, only a minor perturbation of the outline of the capture zone compared to a homogeneous aquifer is expected. As the maximum correlation lengths (see Table 2) used to represent aquifer heterogeneity were significantly smaller than the targeted convergence width of 50 m but still within the same order of magnitude, we expected to model a relatively sensitive part of the real world parameter range. As it was not expected to achieve the same convergence width as in the homogeneous case, a capture width target of 40 m was assumed. This was equivalent to the use of a safety factor of 1.25.

For the geostatistical representation of different anisotropic aquifer heterogeneities the multi-Gaussian field generator "sgsim" from the Geostatistical Software Library (GSLIB, Deutsch & Journel, 1992) was used with an exponential variogram model for unconditional simulations. The angle parameter φ corresponds to a clockwise rotation of the largest principal axis starting from the y-direction of the coordinate system (x-direction corresponds to easting, y-direction corresponds to northing, and z-direction points outwards of the xy-plane according to the right-hand rule).

Cost model

The individual cost elements consisted of the amount of zero valent iron, the sum of all funnel elements, and the sum of all gate elements. The unit prices used are shown in Table 3. It was assumed that a gate was constructed by two slurry wall segments parallel and two sheet pile walls perpendicular to the groundwater flow. The latter were only present during ZVI filling and removed afterwards. Funnels consisted of slurry wall segments.

Table 3 Relevant trichloroethene (TCE) and zero valent iron (ZVI) properties used for the contamination template site.

TCE	$t_{1/2}$ (s)	C_{in}/C_{out} (-)
	37800	0.001
ZVI	ρ_{ZVI} (t m^3)	n_e (-)
	2	0.48

The total costs (C_T) were calculated by the following formula (1):

$$C_T = th \cdot \left[\left(\frac{r \cdot w_{tot}}{1+r} + 2 \cdot \sum_{i=1}^{n_g} l_{g,i} \right) \cdot C_F + 2 \cdot \frac{w_{tot}}{1+r} \cdot C_G + \left(\frac{w_{tot}}{(1+r) \cdot n_g} \cdot \sum_{i=1}^{n_g} l_{g,i} \right) \cdot \rho_{ZVI} \cdot C_{ZVI} \right] \quad (1)$$

The determination of the appropriate reactor or gate length depended on the reactive material and contaminant properties, water chemistry and the actual gate flux Q_i. The contaminant degradation within the reactor was described by a simple first-order model, assuming all contaminant properties were known with certainty. Hence, the required gate length $l_{g,i}$ is only a function of Q_i, the gate width w_g, and aquifer thickness th and can be calculated by the following formula (2):

$$l_{g,i} = \frac{Q_i \cdot \ln\left(\frac{C_{in}}{C_{out}} \right) \cdot t_{1/2}}{w_g \cdot th \cdot n_e \cdot \ln 2} \quad (2)$$

where n_e is the effective porosity of the reactive material, C_{in} represents a constant inflow concentration and C_{out} represents the legal groundwater standard for the particular contaminant and $t_{1/2}$ its half live, respectively. For the calculation of FGS costs a hypothetical trichloroethene (TCE) plume was assumed. The relevant parameter values for the degradation on zero valent iron (ZVI) are presented in Table 4. The uncertainty present in the description of the conductivity field also has an impact on the individual gate fluxes Q_i. Therefore, the different Q_i were recorded for each realisation and grouped to a Q_i–distribution. The 95th percentile was used to compute the necessary gate lengths, $l_{g,i}$ for the cost calculation according to equation (2).

Table 4 Unit costs of the individual cost elements of the applied cost model.

Unit Costs	ZVI-Costs C_{ZVI} (€m^{-3})	Funnel-Costs C_F (€/m^2)	Gate-Costs C_G (€/m^2)
	300	70	190

RESULTS

As an example, a total of 90 different capture zone probability curves were computed. Figure 4 shows a comparison between the FGSs 1G052 and 4G052 for the aquifer heterogeneity scenarios “long” and “short”. For simplicity, results are shown for the central layer only (no. 4). The diagrams clearly depict the dependency of the reliability of capture for a certain w_{cap} on the distance y upstream of the FGS. Even though both designs have the same convergence width, the capture zone reliability curves differ—even for larger distances. In terms of the influence of the direction of the main principal axis of aquifer heterogeneity, it seems that an orientation perpendicular to the FGS elongate axis (sc. “long”) leads to lower reliabilities than in the parallel case (short). In this example the 4-gate FGS appeared to be less sensitive to the aquifer heterogeneity than the single gate FGS, although the relative aquifer heterogeneity (1G052: $\lambda_{max}/w_{tot} = 0.20$; 4G052: $\lambda_{max}/w_{tot} = 0.31$) was larger for the 4-gate system.

Figure 5 shows reliability-cost values for w_{cap} = 40 m of the nine FGS designs for the aquifer scenarios “long”, “short”, “angle”, “iso” and different distances y. In these plots the most favourable designs were found closest to the right and to the bottom of the diagrams. Obviously, no optimal design can be determined for all scenarios and distances, as in a number of cases, the most cost-efficient is not the most reliable and *vice versa*. Considering only the clustering of symbols, it can be inferred that multi-gate systems were more reliable when assessing the ultimate FGS vicinity (y = 10 m) and at larger distances upstream (y = 70 m). Quite conversely, the single gate FGS

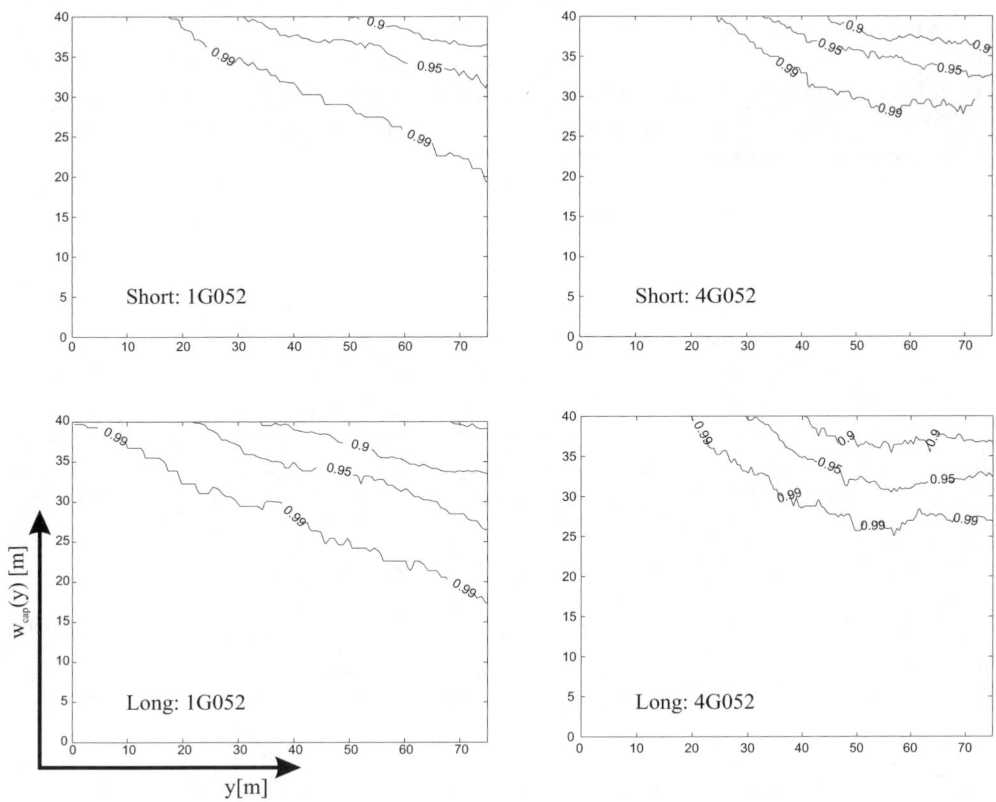

Fig. 4 Capture zone reliability curves of the centre model layer (no. 4) for the aquifer scenarios “long” and “short” and the designs 1G052 and 4G052. Contours represent the reliability of capture (-).

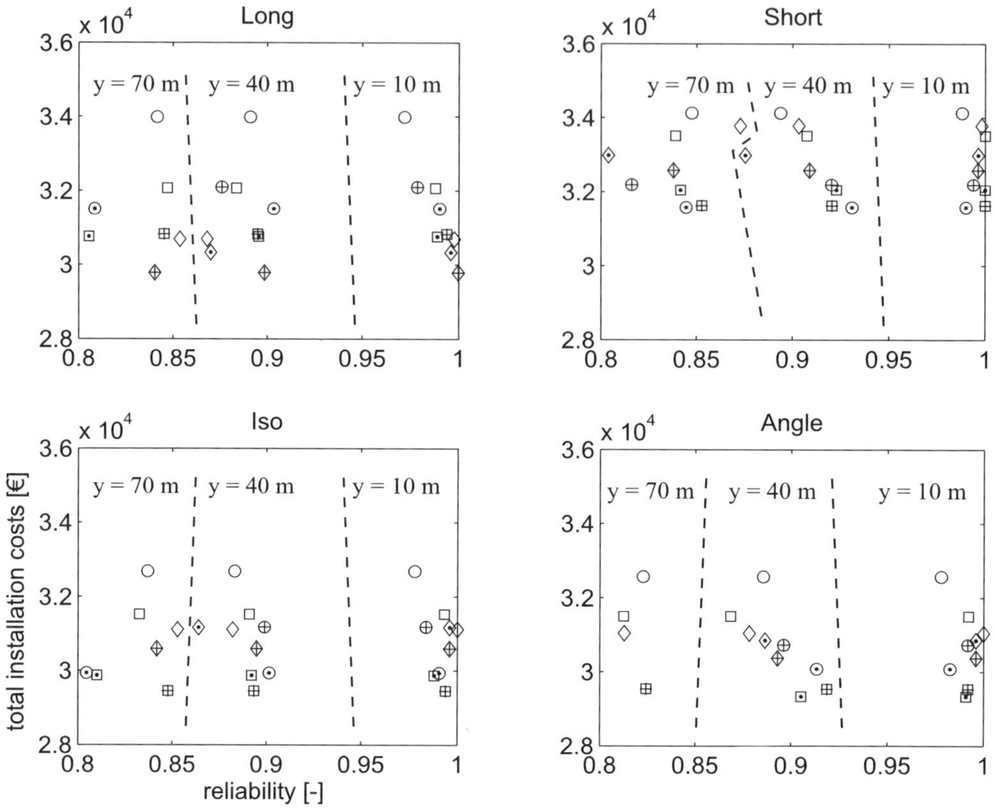

Fig. 5 Reliability–cost plots of the nine funnel-and-gate systems for the aquifer scenarios "long", "short", "iso", and "angle". (Please refer to Figs 6 or 7 for the symbol legend).

1G145 appeared to be most reliable at mid-distances (y = 40 m) for "long", "short" and "iso". A relatively good compromise was found with 2G052, which appeared to be the most cost-efficient and reliable design in the majority of cases. Concerning the influence of the aquifer anisotropy on the order of the FGS designs it appeared that at the mid-point of the design clusters, higher reliabilities at higher costs were found in scenario "short", compared to scenario "long". In scenario "angle" higher reliabilities were found at the mid-point of the design cluster at mid-distance than at the corresponding mid-point for scenario "long". On the other hand, for larger distances, the opposite seemed to be the case.

Figure 6 shows reliability–cost plots for w_{cap} = 40 m for the nine designs and for the aquifer scenarios "v45_long", "v45_short", and "v45_iso" for different distances y. In this extremely heterogeneous case, individual clusters of designs with the same number of gates developed for y = 10 m. Here, multi-gate systems clearly were more reliable than single gate FGS. Nevertheless, in terms of costs, the latter were the most favourable. With further distance y the multi-gate designs came within the same range of reliability as the single gate designs. The overall reliability decreased, as expected, considerably compared to the low variance case. Again, the design 2G052

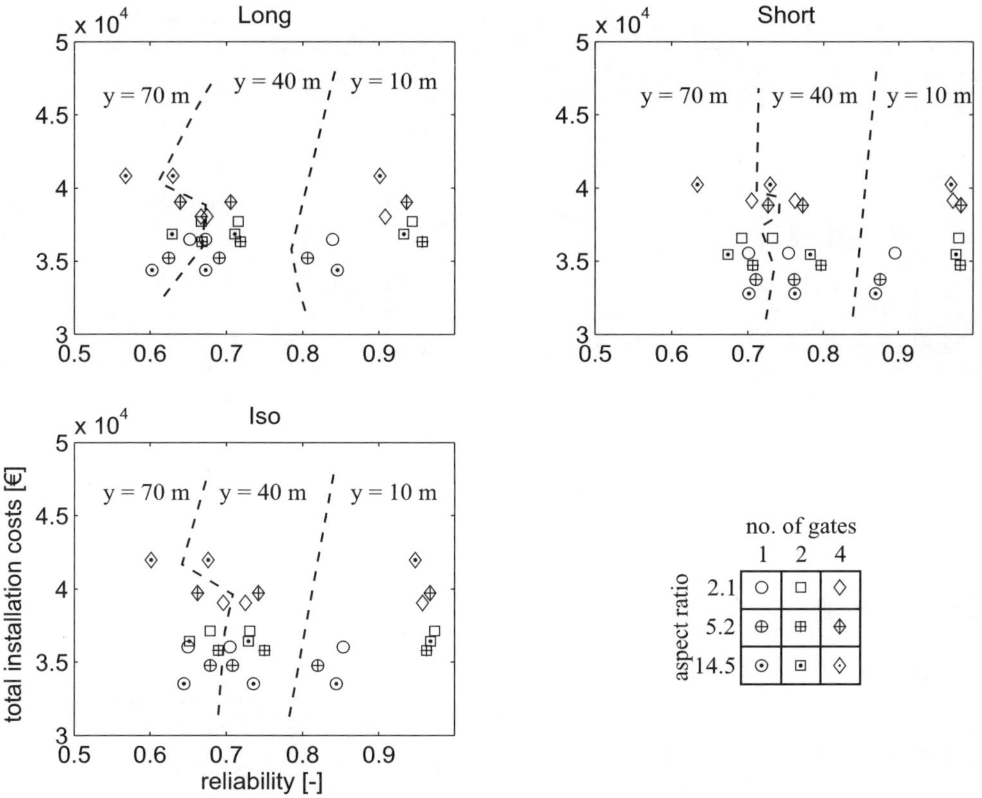

Fig. 6 Reliability–cost plots of the nine funnel-and-gate systems for the aquifer scenarios "v45_long", "v45_short", and "v45_iso" (high variance).

appeared to be a reasonable compromise. Concerning the influence of the aquifer anisotropy on the order of the FGS designs, no clear trends are found regarding the scenarios "v45_long" and "v45_short".

Figure 7 displays the results for aquifer scenarios "d_long", "d_short", "d_iso" with all horizontal correlation lengths doubled. In terms of the overall reliability the expected, more pronounced decrease with distance y was found. Similar to the high variance case, a distinct clustering of single compared to multi-gate designs was noted at $y = 10$ m. Again, the multi-gate designs were more reliable. At further distances, no distinct clustering was discerned, so that the most cost-efficient and reliable was not necessarily attributed to multi-gate designs. Interestingly, design 2G052 was again the most reasonable choice. Concerning the influence of the aquifer anisotropy on the order of the FGS designs, similar findings can be attributed to the scenarios "d_long" and "d_short" as to the associated scenarios "long" and "short".

CONCLUSIONS

The findings presented imply that within a distance of 10 m multi-gate systems reached the targeted capture width of 40 m with reasonably high reliabilities (above

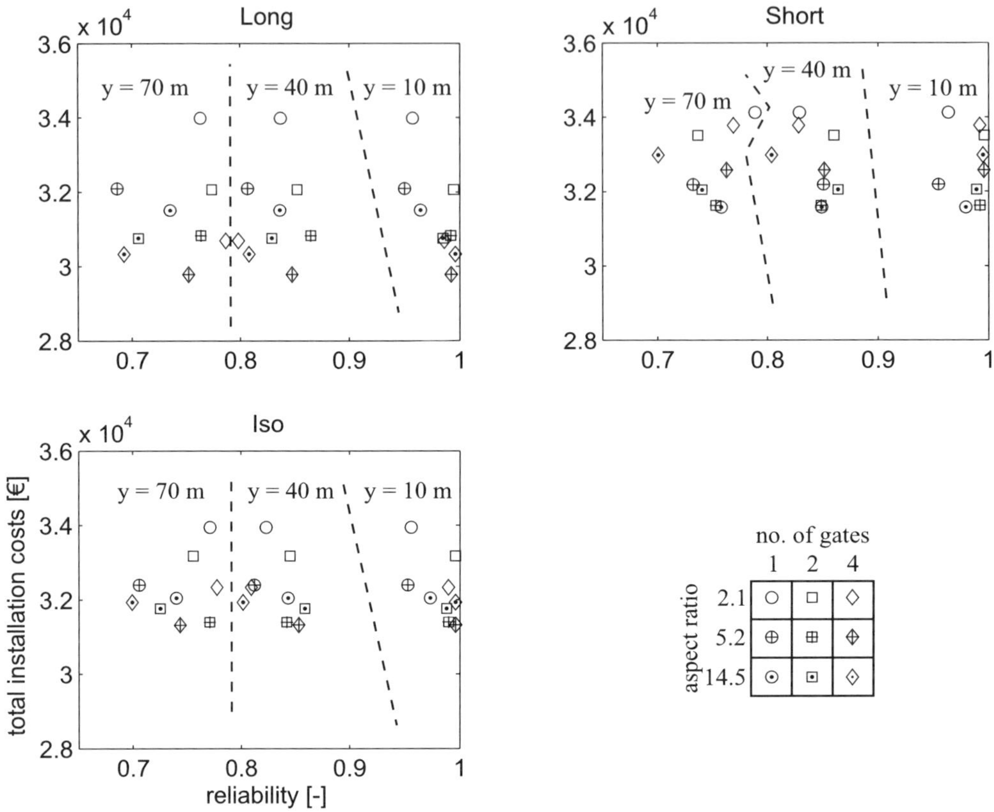

Fig. 7 Reliability-cost plots of the nine funnel-and-gate systems for the aquifer scenarios "d_long", "d_short", and "d_iso" (doubled correlation lengths).

0.95) for all aquifer scenarios. The safety factor applied may be viewed as sufficient for this objective. Nevertheless, with increasing distance y to the FGS, higher safety factors will be needed to obtain similarly high reliabilities. In terms of the overall design cost, the variance of the random conductivity fields seems to play a determining role. In this respect, it may also influence the design selection as single gate systems were the most cost-efficient for all three scenarios with $\sigma^2 = 4.5$. This is most likely due to the fact that in the computation of the gate lengths uncertainty was accounted for n_g times.

For all cases with a lower variance, though, multi-gate systems were among the most reliable and cost-efficient over the whole range of distances y. The findings for the 2G052 system infer that the 2-gate systems are a reasonable compromise in terms of the objectives, cost and reliability. Concerning the design variables aspect ratio and number of gates, the results seemed to show a trade-off rather than a clear trend with respect to the cost and reliability objectives.

Considering the effects of the anisotropy of the hydraulic conductivity, the results give the indication that a relationship of the correlation lengths similar to the "short" scenarios may be more favourable than the one in the 'long' scenarios in terms of system reliability. The findings for the "iso" scenarios seemed to be closer to those for the "long" scenarios than to the "short" ones. The scenario "angle" seemed to have a more negative effect on system reliability than the scenario "long" for longer distances. However, this is expected since the method used for capture width determination relates the capture zone to the symmetry axis of the FGS. As the aquifer anisotropy with an oblique angle to this symmetry axis creates a preferred orientation of the flow field, the expectation of the capture zone is no longer directly related to the symmetry axis. Hence, apart from the finding for the "long"–"short" comparison it appears difficult to identify clear trends. This might be due to the fact that the vertical correlation length λ_z was not changed. As λ_z was the minimum length in all aquifer scenarios it is very likely that a low conductivity zone is rather circumvented by travelling a small vertical distance than a large lateral one.

REFERENCES

Bürger, C., Finkel, M. & Teutsch, G. (2003) Technical and economic evaluation of multiple gate funnel-and-gate systems under homogeneous and heterogeneous aquifer conditions. In: *Calibration and Reliability in Groundwater Modelling: A Few Steps Closer to Reality* (ed. by K. Kovar & Z. Hrkal) (Proc. ModelCARE'2002, June 2002), 448–455. IAHS Publ. 277. IAHS Press, Wallingford, UK.

Dagan, G. (1989) *Flow and Transport in Porous Formations*. Springer-Verlag, Berlin, Germany.

Deutsch, C. V. & Journel, A. G. (1992) *GSLIB: Geostatistical Software Library and User's Guide.* Oxford University Press, New York, USA.

Elder, C. R., Benson, C. H. & Eykholt, G. R. (2002) Effects of heterogeneity on influent and effluent concentrations from horizontal permeable reactive barriers. *Water Resour. Res.* **38**(8), 10.1029/2001WR001259.

Harbaugh, A. W. (1990) A computer program for calculating subregional water budgets using results from the US Geological Survey modular three-dimensional ground-water flow model. *US Geol. Survey, Open-File Report 90-392.*

Harbaugh, A. W., Banta, E. R., Hill, M. C. & McDonald, M. G. (2000) MODFLOW-2000, the US Geological Survey modular ground-water model—User guide to modularization concepts and the ground-water flow process: *US Geol. Survey, Open-File Report 00-92.*

Koltermann, C. & Gorelick, S. (1996) Heterogeneity in sedimentary deposits: A review of structure-imitating, process-imitating, and descriptive approaches. *Water Resour. Res.* **32**(9), 2617–2658.

Pollock, D. W. (1994) User's Guide for MODPATH/MODPATH-PLOT, Version 3: A particle tracking post-processing package for MODFLOW, the U.S. Geological Survey finite-difference ground-water flow model. *US Geol. Survey, Open-File Report 94-464.*

Richards, P. (2002) Construction of a permeable reactive barrier in a residential neighborhood. *Remediation* **autumn 2002**, 65–79.

Teutsch, G., Tolksdorff, J. & Schad, H. (1997) The design of *in-situ* reactive wall systems—A combined hydraulic-geochemical-economical simulation study. *Land Contamination & Reclamation* **5**, 125–130.

Permeable Reactive Barriers (Proceedings of the International Symposium held at Belfast, Northern Ireland, March 2004). IAHS Publ. 298, 2005.

Hydrodynamic modelling for coupled free and porous domains while designing permeable reactive barriers

DIGANTA BHUSAN DAS
Department of Engineering Science, Oxford University, Oxford OX1 3PJ, UK
diganta.das@eng.ox.ac.uk

Abstract Groundwater flow through permeable reactive barriers (PRBs) may often involve coupled free (non-porous) and porous sections. To model the fluid mobility efficiently in such situations and design an effective subsurface remediation scheme, the free and porous flow zones should be studied in conjunction with each other. The flow phenomena in the conjugate regions must also be upscaled from the local to larger scales (e.g. field scale). The present paper focuses on analysing the hydrodynamic conditions in combined free and porous domains in an aid for designing PRBs. The paper does not address the issues related to upscaling the flow processes to a larger scale.

Key words Beavers & Joseph (1967) formulation; combined free and porous flow; Darcy equation; Navier–Stokes equations; permeable reactive barrier

INTRODUCTION

In situ treatment of underground water by permeable reactive barriers (PRBs) has emerged as an important new alternative to the traditional remediation techniques in the last decade. PRBs are *in situ* containment structures that have advanced from an engineering viewpoint in the 1980s to field applications in the last decade. They consist of a reactive porous medium placed *in situ* and downgradient of a contaminant source or plume. Composition of the reactive zones may vary but their hydraulic conductivity (K) must be equal to, or greater than, the surrounding aquifer so that they do not obstruct the groundwater flow. In general, because of the heterogeneity in a PRB and the surrounding aquifer, and the multiplicity of other factors that can affect flow through these regions, prediction of underground flow field characteristics is the most difficult to achieve as the prerequisite in the design of effective reactive barriers (Das, 2002). On the contrary, if the data and the associated uncertainty in the rate and direction of flow and transport in the coupled PRB/aquifer regions are known, the design of an effective PRB can be based on the well-established principles of chemical reaction engineering (see, e.g. Levenspiel, 1972). Despite this, the majority of current studies concerning PRBs have focused on the investigation of the kinetics of geochemical reactions involving different filling materials and their long-term performance while hydrodynamic studies of these systems have remained largely incomplete.

Almost all existing PRBs have been designed assuming the aquifer and the PRBs to be homogeneous or heterogeneous porous media. The design methods generally assume a plug-flow model to estimate residence time of contaminated groundwater in the PRBs and first order reaction kinetics to predict the level of contaminant reduction. A safety factor is then applied to take into account the differences between model assumptions and natural conditions, such as variation in groundwater flow, influent and effluent concentrations in PRBs, uncertainty in media reactivity, difference in laboratory and field conditions; heterogeneity of the aquifer; non-first order reactions, long-term degradation of PRBs, etc. (Elder *et al.*, 2002).

In many cases, sources of underground contamination originate within lands used by industries in the past. Prime examples are the sites of old gas works. As well as the expected heterogeneity in the porous sections (e.g. presence of layers/laminations, micro-scale lenses, etc.), these regions often contain conduits, fissures and other buried structures, which are frequently damaged (Fig. 1). Macro-pores such as vugular inclusions and fractures are especially common in many carbonate rocks and groundwater reservoirs. These regions significantly increase the effective porosity and permeability of the aquifer. Under such circumstances, three mechanisms are expected: (a) the quantity of contaminants transported by convection will be far greater than that transported by diffusion; (b) the time scales associated with convection is significantly reduced and; (c) the dimensions of the convective zones are larger enabling contaminants to spread over much greater distances relative to diffusion. The formation of viscous and/or capillary fingers in a mixed flow regime is also expected. Hence, treating these domains simply as porous media, whether heterogeneous or homogeneous, is not reasonable and may lead to major errors in the prediction of the transport characteristics (Das, 2002; Das, *et al.*, 2002; Das & Nassehi, 2003). Non-porous sections in the subsurface should be generalized as free flow regimes and must be represented in the mathematical models used in the design of PRBs (Das, 2002). Free flow regime can be interpreted as if the flow were in a layer of fluid or a pipe and needs only fluid dynamical characterization of the domain (e.g. Reynolds number). On the contrary, porous flow regime indicates flow through a porous domain and needs characterization for both fluid dynamics and the medium, e.g. permeability, porosity. To the best of the author's knowledge, present design methods for PRBs take no account of coupled free and porous flow scenarios in the subsurface. This is indicated by the fact that no safety factor is prescribed for mixed flow domains. Existing design approaches are unlikely to work for these scenarios and modifications to the current methods must be sought.

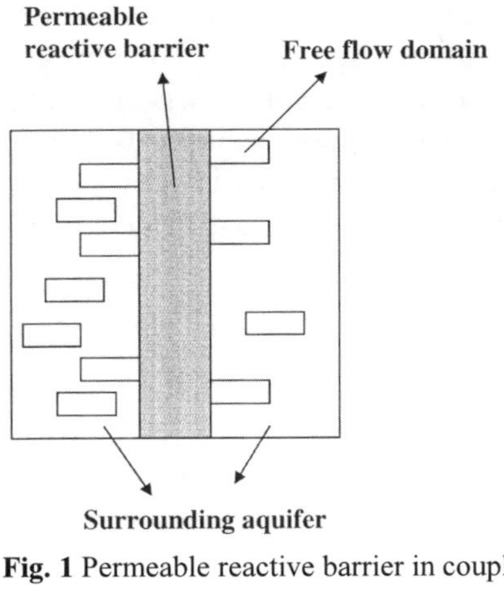

Fig. 1 Permeable reactive barrier in coupled free and porous domains.

Construction of PRBs in coupled free and porous flow regions is not only of importance (e.g. due to the possibility of higher contaminants load) and complexity, but it also needs an integrated approach. The basic hydrodynamics in the combined regions should be resolved before one can move into the contaminant transport problems. This will lead to the problem of chemical and biological kinetics, adsorption/desorption and dissolution kinetics and, interfacial mass transfer in mixed regions. The flow and transport phenomena in coupled flow regions must also be upscaled from the local scale to larger scales, e.g. field scale. This is because the dominant forces and, hence, the form of governing model equations for flow and contaminants transport at the local and larger scales, may be different. A suitable upscaling methodology would identify the appropriate form of governing equations at the larger scale and any coefficient that may arise from averaging the governing equations. Often, one is also interested in determining how the presence of free flow domain at the local scale affects the effective media properties (e.g. hydraulic conductivity) at the field scale. With an upscaling methodology, one can determine the effective media properties of interests. However, we suffer from a lack of accurate enough methodologies for upscaling the local processes to field scale. Hence, we regularly make simplifying assumptions for calculating effective flow and transport properties, which are not necessarily valid at the field scale. To develop an accurate upscaling methodology, and thereby an effective remediation programme, it is necessary to understand and quantify the flow processes at smaller scales (e.g. core scale) and describe their manifestation at larger scales (e.g. field scale). This paper presents an approach for coupling free and porous flow at the core scale without addressing the question of upscaling the processes to a larger scale. Numerical results for various cases are discussed.

Modelling coupled free and porous flow

The present work is based on a two-domain flow model for simultaneous simulation of the hydrodynamics of combined flow between free and porous sections. In this approach, the free and the porous flow domains are considered separately instead of a more common single-domain approach, which considers the porous regime as a pseudo-fluid layer and the whole domain as a continuum. The Navier-Stokes equation (equation (1)) in dimensionless form is adopted as the governing equation for the momentum transfer in the free flowing fluid. The transient form of the Darcy equation in dimensionless form (equation (2)) is used to represent the flow in porous domains. The validity of these equations in the respective cases is well known. Constant fluid density and viscosity at a reference temperature are also assumed.

Navier–Stokes equation:

$$\mathrm{Re}\,\nabla P_f^o = -\rho^o\left[\frac{\partial \mathbf{v}_\mathrm{f}^\mathrm{o}}{\partial t^o} + \mathrm{Re}(\mathbf{v}_\mathrm{f}^\mathrm{o}.\nabla\mathbf{v}_\mathrm{f}^\mathrm{o})\right] + \mu^o\nabla^2\mathbf{v}_\mathrm{f}^\mathrm{o} \tag{1}$$

Darcy's equation:

$$\mathrm{Re}\,\nabla P_p^o = -\rho^o\frac{\partial \mathbf{v}_p^o}{\partial t^o} - \frac{1}{Da}\frac{\mu^o}{\mathbf{K}^o}\mathbf{v}_p^o \tag{2}$$

where, the superscript "*o*" indicates dimensionless variable and, the subscripts "*f*" and "*p*" refer to the free and the porous flow domains. *P* and ***v*** are the pressure and velocity field, respectively, while ρ and μ are, respectively, the constant density and viscosity of the fluid. Re and Da are normalized Reynolds and Darcy numbers. **K** is the permeability tensor of the porous media. The conservation of mass in both free and permeable domain is represented by the continuity equations as:

$$\nabla . v_e^o = 0 \quad e = f, p \tag{3–4}$$

As the matching interfacial condition for linking the equations of motion, a modified form of Beavers & Joseph (1967) formulation is used (Das, 2002; Das & Nassehi, 2003). When flow in the porous domain is governed by the Darcy equation, an empirical interfacial slip-flow boundary condition proposed by Beavers and Joseph can be used, which describes proportionality between the shear rate at the interface and the slip velocity through a dimensionless slip coefficient. The dimensionless coefficient depends only on the structural properties of the interface. As adopted in this work, the interfacial boundary conditions at the free/porous interface are (Das, 2002):
Longitudinal velocity component, *u*:

$$u_f^o = u_p^o \tag{5}$$

Lateral velocity component, *v*:

$$\left(\frac{\partial v^o}{\partial x^o} + \frac{\partial u^o}{\partial y^o}\right)_f = \frac{\gamma_{yz}}{\sqrt{DaK_{yy}^o}}\left(v_f^o - v_p^o\right) \tag{6}$$

Transverse velocity component, *w*:

$$\left(\frac{\partial w^o}{\partial x^o} + \frac{\partial u^o}{\partial z^o}\right)_f = \frac{\gamma_{yz}}{\sqrt{DaK_{zz}^o}}\left(w_f^o - w_p^o\right) \tag{7}$$

Pressure, *P*:

$$\left(-P^o + 2\mu \frac{\partial u^o}{\partial x^o}\right)_f = -P_p^o \tag{8}$$

where, γ_{yz} is a non-dimensional slip coefficient at the interface (*y*–*z* plane in this case) and K_{yy} and K_{zz} are the components of the permeability tensor in the lateral (*y*-) and vertical (*z*-) co-ordinate directions.

Numerical experiments

The solution algorithm used in this work comprises a number of important steps and the imposition of initial and boundary conditions. Most salient features of the algorithm have been presented elsewhere (Das & Nassehi, 2001; Das, 2002; Das *et al.*, 2002) and are not discussed in length in this paper. The partial differential equations used in the simulation are discretized and reduced to algebraic forms using the standard finite volume method (FVM). This method is adopted due to its inherent capability in coping with complex 3-D domain types. The simulator calculates the flow variables on a 3-D forward-staggered grid where cell faces are ahead of cell nodes. The mesh for both fluid and porous regions involve rectangular or cubicle cells for storing the flow variables (see, Das & Nassehi, 2001). While the velocity components are evaluated at different cell faces, pressure terms are predicted at cell nodes such that they act as the driving potential for flow at intermediate cell faces. Das *et al*. (2002) have derived the non-dimensional working equations using the FVM and discussed the convergence of numerical solutions through mesh refinements. This work uses the same numerical mesh, which involves 5832 cells for each sub-domain. Both free/porous and porous/porous domain interfaces consist of 324 cells. The dependence of the numerical solutions on the chosen time steps has been examined by Das & Nassehi (2003). To obtain the desired algebraic working equations, the governing partial differential equations are integrated over a cell volume, ΔV, and then discretized using the central differencing scheme for spatial discretization and the explicit scheme for temporal discretization, respectively. Another key point, i.e. the imposition of boundary and initial conditions, must be addressed briefly. The study uses Dirichlet-type boundary conditions at the inlet boundaries of the computational domain and for specifying the initial conditions in both sub-domains. At the sidewalls, non-slip conditions for velocity are imposed. Non-zero initial conditions for velocity are assumed for both the groundwater zone and PRBs. The initial conditions are further assumed to be the same in both sub-domains. At the exit of the porous section, stress-free boundary conditions are imposed as they enable the most realistic simulation of the flow processes.

DISCUSSION

The hydrodynamic conditions for coupled flow through PRBs will be influenced by a large number of factors. Therefore, a detailed general analysis of the problem cannot be done in a single paper. Das (2001) has discussed four approaches for combining the free and porous flow zones in the subsurface and concluded that not every technique is applicable for modelling such mixed flow regimes. Although superficial similarities may be observed in the mathematical models of different coupled flow problems in natural hydro-environmental systems and other chemical and mechanical engineering problems, the combined flow behaviour in the subsurface may differ from those flow processes (Das, 2002; Das *et al*., 2002). These differences are generally brought about by the geometry, structure, the orientation of the free/porous interface(s) and boundary conditions. Another aspect, namely whether the problems involve a moving or constant boundary is of considerable practical significance. Numerical results for transient coupled flow (Das, 2002; Das *et al*, 2002) have shown that at the open end of the porous domain (PRB in this case), co-existing inflow and outflow may take place (Fig. 2(a),(b)). In other words, groundwater from the surrounding aquifer may flow into the PRB through the open end besides through the defined inlet (Fig. 2(b)). For the particular domain investigated, while flow in the free flow domain is mostly uni-directional towards the fluid/porous interface (Fig. 2(a)), the flow may circulate in the porous section and reverse direction at the free/porous domain interface. Evidence of flow circulation, consistent with the numerical results from Das *et al.*, has been given in a number of previous studies representing various scenarios of subsurface flow (e.g. Devito *et al.*, 1997).

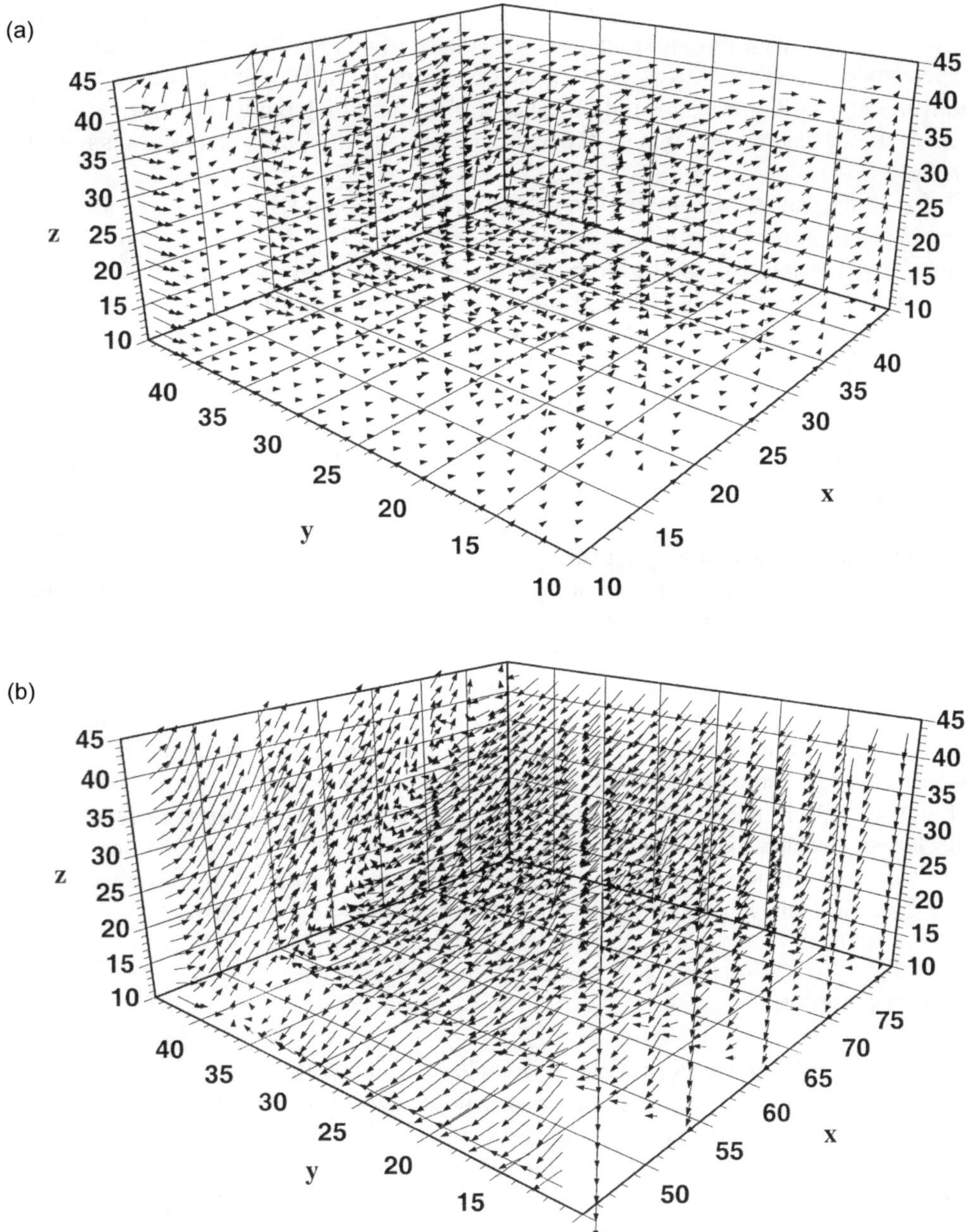

Fig. 2 Sample 3-D velocity field in: (a) the free-flow region (dimensionless) and (b) in the porous-flow region (dimensionless). Velocity vectors are plotted against (x,y,z) co-ordinates. The free/porous domain interface is at (45,y,z).

The fluid circulation in the porous side of the coupled flow domain is a complex transient process. This is generally indicated by the movement of a centre of circulation and a front of flow reversal from the free/porous interface. After a certain time interval (which may depend on many factors including domain characteristics) the flow circulation may disappear and the porous flow may become unidirectional towards the exit of the porous section. The main cause of this complex fluid flow pattern is attributed to variable pressure distributions in the porous domain. The dependence of possible flow circulation on the material properties has largely been unexplored in the previous studies. The implications of having heterogeneity in the porous media properties, such as expected in PRBs, on the flow behaviour also need an in-depth investigation.

In many cases, the interfaces between the free and porous sections can be considered as impermeable for all practical purposes. This is analogous to a scenario where the permeability of the PRBs reduces to such a level that it is lower than the surrounding aquifer. In that case, groundwater may not enter the PRB and may bypass it instead. Das & Nassehi (2001) have analysed a similar problem where the interface has been defined as impermeable. In this problem, the dominant direction of the by-passed flow at the free/porous domain interface has been found to be in the vertical direction. Although these conclusions may not be universally correct as the numerical results depend on the imposed initial and boundary conditions, they give an indication of the discernable effects that may occur if permeability of the PRB becomes less than that of the surrounding media. In such cases, contaminants may not enter the PRBs and may move to the surrounding aquifer.

The length scales of free and porous domains are also important factors in determining the coupled flow behaviour. Das & Nassehi (2002) have investigated the scale effects on the hydrodynamics of coupled flow by choosing different aspect ratios for the sub-domains and showed that they affect the flow circulation in porous domain. The consequence of having more than one fluid/porous interfaces in the flow domain on the concentration profiles of a conservative pollutant has been analysed by Das & Nassehi (2003). They have indicated that, in general, flow fields in free flow domains are not strongly influenced by the presence of the permeable interfaces except near the interfaces. The dominant directions for contaminants flow in the porous domain may vary from one domain to another, depending on the orientation of the interface, among other factors such as material properties and direction of flow gradient.

CONCLUSIONS

The present paper aims to propose a modelling framework, which may be used for design of PRBs in coupled free and porous flow regimes in the subsurface. The fluid under the ideal field conditions can move from free flow zone to porous flow zone, or *vice versa*, depending on the direction of the flow potentials across the domains. Further, the domain may have one or a multiple number of interfaces across which flow can take place. The free flow zone has infinite permeability and needs only fluid dynamical characterization for the mathematical formulation. On the other hand, the porous flow region is subjected to the additional constraints of characterizing the media, e.g. the finite permeability of the porous medium, apart from the necessary flow description. Knowledge of the mass and momentum transfer across the free/porous flow domain interface is crucial because it accounts for the transition in flow type and act as the inlet boundary conditions for the flow domain ahead. Therefore, the mathematical model for the conjugate transport regimes in the subsurface zones should be designed in such a way that it can represent the flow and transport behaviour in both domains together with the transition (jump) of flow variables, if any, across the interfacial plane(s). The goal may be achieved in a number of ways; however, one must be careful that it is suitable to the problem in hand. The application of an approach for the design of PRBs would require upscaling the coupled flow behaviour from the smaller scales to larger scales.

REFERENCES

Beavers, G. S. & Joseph, D. D. (1967) Boundary conditions at naturally permeable wall. *J. Fluid Mechanics* **30**(1), 197–207.

Devito, K. J., Waddington, J. M. & Branfireun, B. A. (1997) Flow reversal in peatlands influenced by local groundwater systems. *Hydrol. Processes* **11**, 103–110.

Das, D. B. (2001) Computer modelling of contaminants mobility in subsurface regions. PhD Thesis, Loughborough University, UK.

Das, D. B. (2002) Hydrodynamic modelling for groundwater flow through permeable reactive barriers. *Hydrol. Processes* **16**, 3393–3418.

Das, D. B. & Nassehi, V. (2001) LANDFLOW: A finite volume model of combined free and porous flow of water in contaminated land sites. *Water Sci. Technol.* **43**(7), 55–64.

Das, D. B. & Nassehi, V. (2002) Development of a new mathematical model for subsurface water quality management. *Water Sci. Technol.* **45**(9), 301–307.

Das, D. B. & Nassehi, V. (2003) Influence of multiple interfacial boundaries on contaminants' mobility in subsurface regions. *Water Resour. Res.* **39**(3), 1072.

Das, D. B., Nassehi, V. & Wakeman, R. J. (2002) A finite volume model for the hydrodynamics of combined free and porous flow in sub-surface regions. *Adv. Environ. Res.* **7**(1), 35–58.

Elder, C. R., Benson, C. H. & Eykholt, G. R. (2002) Effects of heterogeneity on influent and effluent concentrations from horizontal permeable reactive barriers. *Water Resour. Res* **38**(8), art. no. 1152.

Levenspiel, O. (1972) *Chemical Reaction Engineering*. Wiley, New York, USA.

Permeable Reactive Barriers (Proceedings of the International Symposium held at Belfast, Northern Ireland, March 2004). IAHS Publ. 298, 2005.

Evaluation of the hydraulic performance of a full-scale permeable reactive barrier by a tracer study

KEITH C. K. LAI[1], PETER KJELDSEN[2] & IRENE M. C. LO[1]

1 *Department of Civil Engineering, Hong Kong University of Science and Technology, Clear Water Bay, Kowloon, Hong Kong*
cemclo@ust.hk

2 *Environment & Resources DTU, Technical University of Denmark, Lyngby, Denmark*

Abstract Investigation of the hydraulic performance of a funnel-and-gate permeable reactive barrier (PRB), packed with zero-valent iron (Fe^0) and installed at the Vapokon site, Denmark, has been conducted with a natural gradient tracer study. The purposes of this study were to evaluate the flow pattern and determine the seepage velocity (v_x) of groundwater through the PRB. After collecting and analysing about 13 000 groundwater samples over a period of 10 months, the moving path of the tracer (i.e. lithium, Li^+) was identified in which the Li^+ plume was observed passing through the reactive barrier. However, probably owing to clogging caused by mineral precipitates, there was a preferential path inside the Fe^0 PRB. Comparison of the water table contour in the Vapokon site obtained in March 2000 and January 2003 showed formation of a low permeability zone within the reactive barrier, thereby further verifying clogging caused by mineral precipitates. Spatial moments analysis of the Li^+ distribution illustrates the similarity of the maximum Li^+ mass passing through the upgradient and downgradient locations of the Fe^0 PRB and thereby indicated the conservative of the Li^+ across the reactive barrier as well as the absence of bypassing flow. Based upon the results of first moment analysis, a v_x of 99.5 m year^{-1} within the PRB was calculated. The v_x just upgradient of the reactive barrier, however, was only about 6.86 m year^{-1}, most likely owing to the effect of disturbance exerted by the clogged upgradient interface of the upper part of the Fe^0 PRB.

Key words hydraulic performance; *in situ* tracer study; permeable reactive barrier

INTRODUCTION

Permeable reactive barrier (PRB) is a state-of-the-art approach for *in situ* remediation of groundwater contaminated by chlorinated aliphatic hydrocarbons (CAHs) such as trichloroethylene (TCE) and tetrachloroethylene (PCE) (Puls *et al*., 1999). By definition, it involves an emplacement of reactive material in the subsurface to intercept a contaminant plume in which toxic contaminants are transformed into benign products (USEPA, 1998). Since contaminated groundwater flows through a PRB passively by its natural gradient, *in lieu* of mechanical pumping, it is more cost-effective than the conventional pump-and-treat system and has thereby recently drawn much attention from the public (Scherer *et al*., 2000). Among varied reactive materials packed in a PRB, 29 out of 53 full- and pilot-scale PRBs applied zero-valent iron (Fe^0) as reactive medium because of its low cost and high reactivity (Gillham *et al*., 1994; USEPA, 2002).

Nowadays, intensive monitoring of the performance of full-scale Fe^0 PRBs has been widely carried out, such as the first barrier installed at Canadian Forces Base (CFB), Borden, Ontario, so as to investigate the long-term reactivity of Fe^0 on dechlorinating

CAHs from contaminated groundwater (Reynolds *et al.*, 2002). However, calculation of the rate of dechlorination by Fe^0 cannot be achieved without understanding the groundwater seepage velocity flowing through the reactive barrier. Besides, the flow pattern of groundwater through a PRB is rarely investigated in spite of its significance for assessing whether contaminated groundwater is bypassing a PRB or a PRB has any change in its permeability (Focht *et al.*, 1997). Virtually, based upon the observations from laboratory studies, there is a high likelihood of precipitation of calcite ($CaCO_3$), siderite ($FeCO_3$) and metal sulphide (MeS) on the Fe^0 surface (USEPA, 1998), which potentially can reduce the void space of the Fe^0 PRB and thereby affect the barrier's permeability. For the sake of understanding the seepage velocity (v_x) as well as flow pattern of groundwater through a Fe^0 PRB, a natural gradient tracer study using lithium ion (Li^+) as a tracer was conducted for the funnel-and-gate Fe^0 PRB installed at Vapokon, Denmark, between November 2002 and August 2003.

FIELD DESCRIPTION

The Vapokon site located in the northern part (Søndersø) of Fuen Island of Denmark was polluted by a solvent recycling factory. The leakage of solvent mainly containing 1,1,1-trichloroethane (1,1,1-TCA), TCE, PCE and aromatic organics (BTEX) from underground storage tanks subsequently contaminated the groundwater and created a 200×80 m contaminant plume. In light of the recalcitrant and carcinogenic propensities of CAHs (Lai *et al.*, 2002), a funnel-and-gate Fe^0 PRB consisted of two sheet pilings (funnel) with a length of 110–130 m and a Fe^0 barrier (gate) with a dimension of $14.5 \times 9 \times 0.8$ m (W × D × L) was constructed in September 1999 for groundwater remediation (Fig. 1). In the locations just upgradient and downgradient of the Fe^0 PRB, gravel packs with a thickness of 1 m were installed so as to enhance the mixing of contaminated groundwater. Prior to the installation of the Fe^0 PRB, the natural v_x of the Vapokon site was about 400–500 m $year^{-1}$. However, the drainage system installed in the upgradient of the funnel-and-gate Fe^0 PRB prevents any uncontaminated groundwater from flowing into the capture zone of the funnel-and-gate system, thereby greatly reducing the v_x flowing through the PRB.

TRACER STUDY AND FIELD MONITORING

Apart from the 15 monitoring wells (three screens per each well) originally installed across the Fe^0 barrier during the time of PRB construction, a dense network of multi-level sampler (MLS) consisting of six copper MLSs (42 screens) and 20 nylon MLSs (280 screens) was also installed thereafter by using Geoprobe technique for the processing of the tracer experiment (Fig 2(a),(b)). Copper MLS surrounded by a hollow iron pipe was installed within the PRB since the Fe^0 was found collapsing poorly around the MLSs, whereas in the upgradient and downgradient location of the Fe^0 PRB, nylon MLS was installed because of complete collapse of the sediment around the MLSs. However, probably owing to the mechanical damage of the copper sampling tubes during the installation process, 12 out of 42 screens of the copper MLSs malfunctioned.

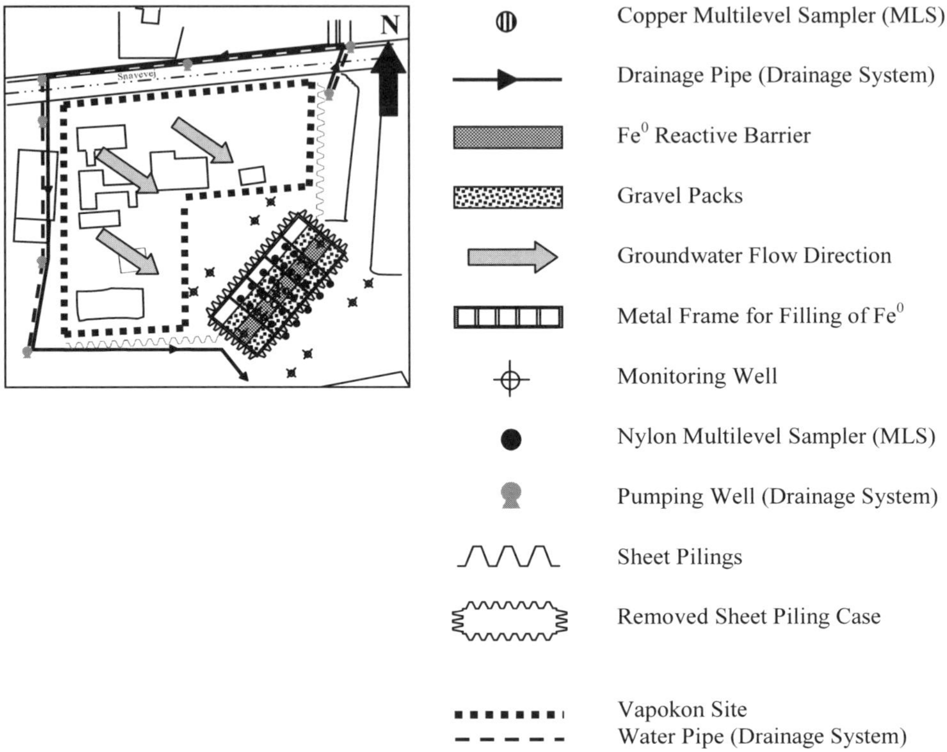

Fig. 1 A funnel-and-gate Fe^0 PRB installed at the Vapokon Site (not to scale).

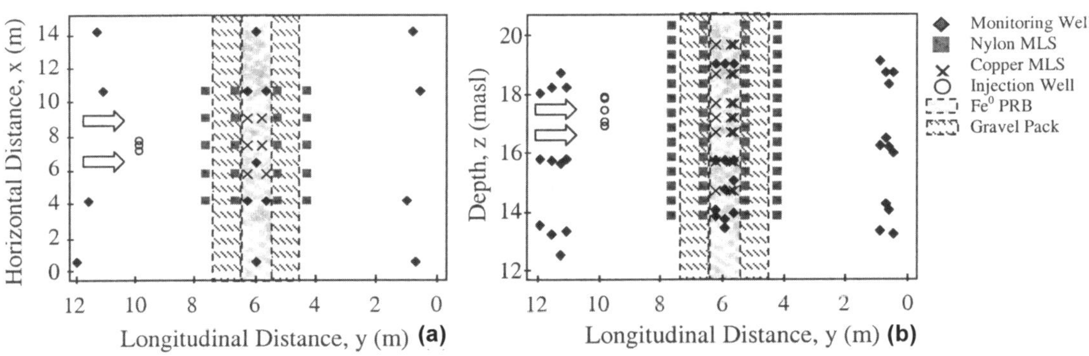

Fig. 2 (a) Planar and (b) side distributions of the monitoring wells, injection wells, nylon and copper MLSs, across the Fe^0 PRB.

The copper MLS consisted of a hollow iron pipe (OD, 28 mm; ID, 14 mm) with seven copper sampling tubes (OD, 5 mm; ID, 3 mm) attached inside with a vertical spacing of 0.5 or 1 m (Fig. 3(a)). Every copper tube was screened by connecting to the slotted opening on the hollow iron pipe, which has a length of 7.5 cm. In contrast, the nylon MLS consisted of a hollow PVC pipe (OD, 20 mm; ID, 10 mm) with 14 nylon sampling tubes (OD, 4 mm; ID, 2.5 mm) attached outside with a vertical spacing of

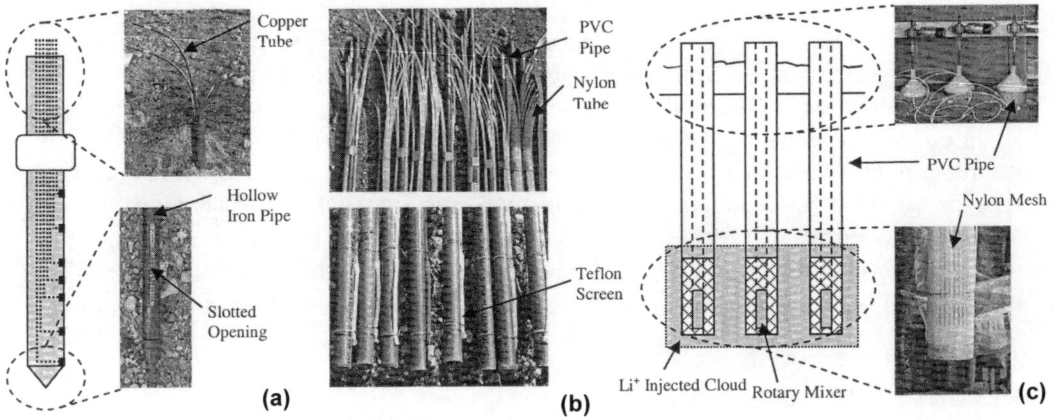

Fig. 3 (a) Copper MLSs, (b) nylon MLSs, and (c) injection wells.

0.5 m (Fig. 3(b)). Each nylon tube was screened by connecting to a slotted Teflon tube (OD, 4 mm; ID, 3 mm) with a length of 7.5 cm.

Li^+ was injected into the three injection wells installed approximately 3.5 m up-gradient of the Fe^0 PRB by using a sediment bailer and steel casing. Each injection well consisted of a PVC pipe (OD, 125 mm; ID, 117.5 mm) screened from 16.8 to 17.8 m above sea level (m a.s.l.) or 4.4 to 5.4 m below ground surface and wrapped with a nylon mesh (Fig. 3(c)). The three injection wells were located 0.275 m (centre to centre) apart on a line perpendicular to the groundwater flow direction, thus producing an injected cloud with an initial width of about 0.81 m. Groundwater passed through the well screens by the natural gradient controlling the cloud movement. Three vertical sampling points locating in the upper, middle and lower positions of the well screens were placed on each injection well to monitor the homogeneity of the injected cloud (Broholm *et al.*, 2001).

Li^+ was selected as tracer because of its proven nonreactivity in the Fe^0 packed medium as well as the low background concentration in Vapokon groundwater. Approximately 127 l of feeding solution prepared using the Vapokon groundwater and containing about 3000 mg l^{-1} of Li^+ was continuously injected into the saturated zone in the first 61 days of the tracer experiment. The injection system, which was controlled by a programmable logic controller, sequentially injected 1.44 ml of feeding solution over 60 s once every 180 s in each of the three injection wells. The rotary mixer was operated continuously during the injection period to ensure a complete mixing in each well. In order to minimize the disturbance to natural groundwater flow, the injection rate of the feeding solution was designed so that the volume of injected solution was always <2% (a dilution factor of >50) of the volume of natural groundwater flowing through the injection wells (Broholm *et al.*, 2001).

Continuous injection of Li^+ was started on 1 November 2002 and ended on 1 January 2003. Within this period, groundwater sampling was conducted every 1–4 days. After ceasing the injection, the sampling frequency reduced gradually. Prior to the collection of groundwater samples, 50 ml and 5–15 l of groundwater were first removed from the MLSs and monitoring wells, respectively, so as to prevent the collection of stagnant water. Collected groundwater samples were immediately stored in 20-ml PE bottles with screw caps. After undergoing 4 h of gravitational settling, the supernatant was extracted from the PE bottle for Li^+ measurement with a flame photometer (PEP 7, *Jenway*). In total, groundwater samplings were conducted 70 times over a period of 280 days. As a result, breakthrough curves (BTCs) from 48 discrete sampling points were obtained. In addition, determinations of pH, redox potential (Eh), specific conductance, calcium ion (Ca^{2+}), total iron ion (Fe_T), total alkalinity (TAL) and sulphate ion (SO_4^{2-}) from the collected groundwater samples were also conducted in September 2002 and January 2003 in order to understand how the geochemical behaviour of the Fe^0 PRB affects the flow pattern of groundwater passing through it.

RESULTS AND DISCUSSION

Flow pattern of groundwater through the funnel-and-gate Fe^0 PRB

During the injection period, an injected cloud with a Li^+ concentration of about 220 mg l^{-1} (a dilution factor of ~14) was created around the injection wells. However, mainly due to the dilution by groundwater, a maximum Li^+ concentration of approx. 50 mg l^{-1} could only be measured in the downgradient area. The planar and side views of the movement of the Li^+ plume across the Fe^0 PRB are illustrated in Fig. 4(a) and (b), respectively. As shown in Fig. 4(a), before entering the Fe^0 PRB, the Li^+ plume moved exactly along the centreline of the PRB without pronounced lateral displacement since the three injection wells were also situated along the centreline. However, probably due to the clogging of the central or northeast parts of the Fe^0 PRB caused by mineral precipitates, there was a sudden change of the flow path in the immediate vicinity of the front edge of the Fe^0 PRB. Based upon the results observed on Days 40 and 84, within the Fe^0 PRB, the Li^+ plume moved preferentially along the path which was about 1.5 m apart from the southwest side of the centreline.

Figure 4(b) shows the vertical distribution of the Li^+ plume across the PRB. The Li^+ plume was first detected from the sampling point located just below the water table (i.e. at an altitude of about 20.3 m a.s.l.) on Day 15. Since the injection wells were screened from 16.8 to 17.8 m a.s.l., an upward movement of the Li^+ plume from the injection wells to the front edge of the upgradient gravel pack was expected. This upward movement of the Li^+ plume is most likely ascribed to the presence of a clay lens between the injection wells and upgradient gravel pack, since clay soil could be collected from a depth of 16–19 m a.s.l. during the installation of injection wells. It is believed that this clay lens hindered the Li^+ plume from moving towards the Fe^0 barrier in the altitude, the same as those of the well screens, and thereby forced the upward movement of the Li^+ plume. However, within the upgradient gravel pack, the Li^+ plume dropped significantly from an altitude of about 20.3 to nearly 16.5 m a.s.l. Once the Li^+ plume entered the Fe^0 PRB, it travelled at a depth between 16 and 18 m a.s.l. with a minor drop of altitude. Such an insignificant drop of the altitude within the PRB was probably due to the influences of the density contrast between the Li^+ plume and native groundwater, as observed at Borden, Ontario, Canada (Mackay *et al.* 1986) and Cape Cod, Massachusetts, USA (LeBlanc *et al.*, 1991) as well as rainfall

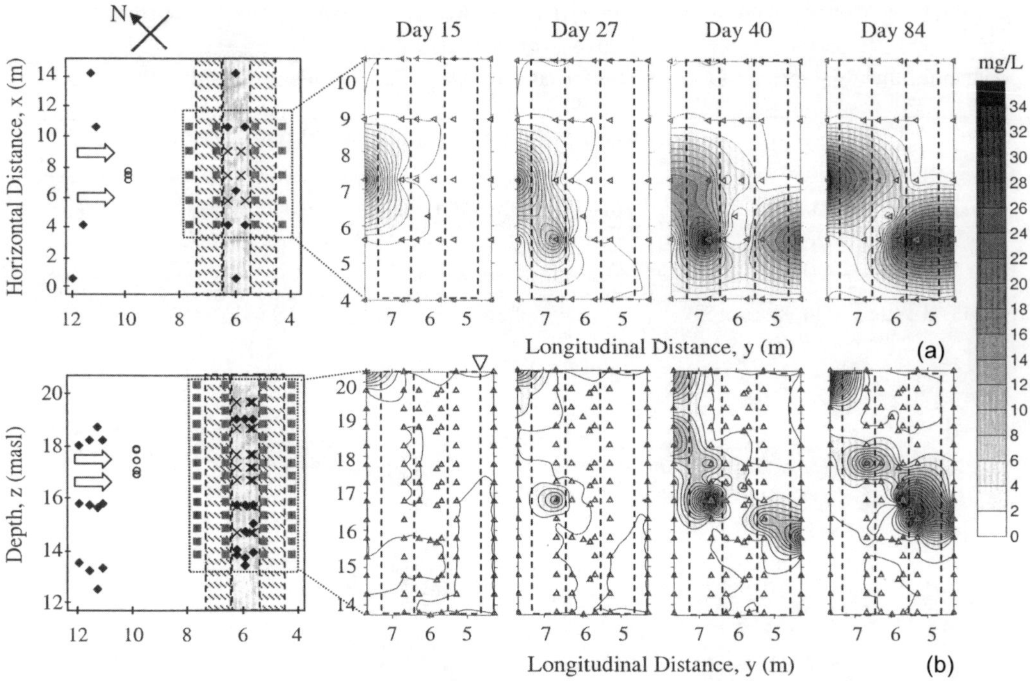

Fig. 4 (a) Planar and (b) side views of the movement of the Li^+ plume across the Fe^0 PRB.

precipitation, which added a layer of water on the groundwater surface and thereby suppressed the altitude of the Li^+ plume. The significant sinking of the Li^+ plume within the upgradient gravel pack, however, was not merely caused by the influences of the density contrast and rainfall precipitation. It is believed that the clogging of the upper part of the Fe^0 barrier caused by mineral precipitates was the main underlying reason, which prevented the Li^+ plume from entering the upper part of the Fe^0 PRB and consequently forced the downward movement.

The preferential path for the Li^+ plume travelling across the Fe^0 PRB is most likely induced by the mineral precipitates formed inside the Fe^0 PRB. Previous study of the geochemical behaviour of the Vapokon Fe^0 PRB had verified that there were significant reductions in the concentrations of Ca^{2+} (79.4%), magnesium (Mg^{2+}, 56.9%), Fe_T (96.2%) and TAL (92.7%) in the groundwater across the PRB (Lo *et al.*, 2003). Similar phenomena could also be attained through the physical and inorganic analyses of the groundwater samples collected in September 2002 and January 2003 (see Table 1). According to Table 1, high pH and low Eh conditions were observed within the PRB, which provided a thermodynamically favourable condition for the precipitation of Ca^{2+}, Fe_T, TAL and SO_4^{2-}, thereby leading a diminution in specific conductance in the groundwater across the Fe^0 PRB. Among varied mineral precipitates, $CaCO_3$, $FeCO_3$, MeS and carbonate-containing green rust ($[Fe_4^{II}\ Fe_2^{III}(OH)_{12}][CO_3{\cdot}2H_2O]$) are most likely formed inside the PRB (USEPA, 1998).

Further verification of the influence of mineral precipitation on the groundwater flow pattern through the Fe^0 PRB can also be obtained by examining the change of the

Table 1 Variation of the physical and inorganic characteristics of Vapokon groundwater along the Fe^0 PRB.

Sampling time/ parameters	pH	Eh (mV)	Specific conductance ($\mu S\ cm^{-1}$)	Ca^{2+} ($mg\ l^{-1}$)	Fe_T ($mg\ l^{-1}$)	TAL ($mg\ l^{-1}$ as $CaCO_3$)	SO_4^{2-} ($mg\ l^{-1}$)
September 2002	na	na	na	147	0.165	na	29.5
				54.3	0.124		3.70
				19.3	0.051		1.75
January 2003	7.16	105.2	646	102	0.070	255	na
	9.44	-101.3	309	38.4	0.107	125	
	10.2	-150.2	229	19.3	0.263	35.4	

Note: The upper, middle and lower figures indicate the average value obtained from the front, middle and end of the Fe^0 PRB, respectively, for the sampling. na, not available.

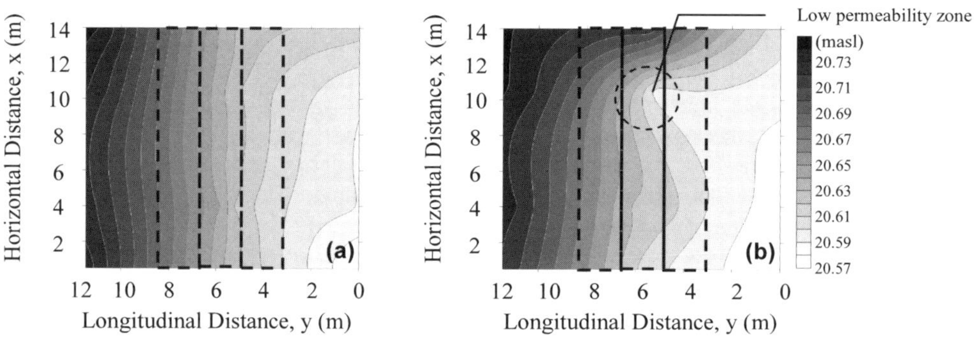

Fig. 5 Water table contour measured in: (a) March 2000 and (b) January 2003.

water table contour at the Vapokon site. As shown in Fig. 5(a), the water table distribution across the Fe^0 PRB was fairly uniform in March 2000. However, in January 2003 (see Fig. 5(b)), a low permeability zone was formed inside the PRB probably owing to the clogging caused by mineral precipitates.

Groundwater seepage velocity within the funnel-and-gate Fe^0 PRB

The spatial moments of the Li^+ distribution were used to compute the Li^+ mass (zeroth moment) and velocity (first moment) across the reactive barrier during the natural gradient tracer study (Aris, 1956). Estimates of the spatial moments were conducted by using a numerical integration of the Li^+ distribution in which the values of concentration at the sampling points were linearly interpolated over the domain and then integrated to obtain the moment estimates (Garabedian *et al.*, 1991). For this analysis, the 3-D spatial moments of the following form were used:

$$M_{ijk} = \iiint_{\Omega} \left(x - u\bar{x}\right)^i \left(y - u\bar{y}\right)^j \left(z - u\bar{z}\right)^k nc \, \mathrm{d}x \, \mathrm{d}y \, \mathrm{d}z \tag{1}$$

where c is the solute concentration ($mg\ l^{-1}$); i, j and k are exponents with values 1 and 2; n is porosity; u is a flag for the central moment with value 0 or 1, depending on the moment being evaluated; x, y and z are the coordinate directions; $\bar{x}$, $\bar{y}$ and $\bar{z}$ are the first moment of the Li^+ distribution; Ω is the test domain.

Since the solute concentration was sampled at points rather than continuously in space, it was required to interpolate the solute concentrations as part of numerical integration procedure. Integration over the vertical was the first step to be performed at each sampler location because the horizontal positions are the same for each set of concentrations at a MLS. The concentration was hypothesized varying linearly between each pair of adjacent sampling points in the vertical. Afterwards, integration over the horizontal plane was conducted in which a linear interpolation between three sampler locations (i.e. a triangular subdomain) of the vertically integrated values was assumed and substituted into equation (1), thereby creating equation (2):

$$M_{ijk} = \sum_{p=1}^{N} \iint_{\Omega_p} n(A + Bx + Cy)_k \left(x - u\bar{x}\right)^i \left(y - u\bar{y}\right)^j \, \mathrm{d}x \, \mathrm{d}y \qquad (2)$$

where A, B and C are the coefficients defining the vertically integrated concentration distribution; k is the exponent of the vertical spatial integration; N is the number of triangular subdomains in the tracer test domain (Ω); Ω_p is the triangular subdomain.

A set of equations from finite element analysis were used to solve equation (2). In fact, each of the expressions within parentheses in equation (2) is a linear function over the triangular subdomain and thereby can be represented as a sum of linear basis functions, resulting in equation (3):

$$M_{ijk} = \sum_{p=1}^{N} \iint_{\Omega_p} n\left(\sum_{L=1}^{3} I_L \phi_L\right)_k \left(\sum_{L=1}^{3} \left(x - u\bar{x}\right)_L \phi_L\right)^i \left(\sum_{L=1}^{3} \left(y - u\bar{y}\right)_L \phi_L\right)^j \mathrm{d}x \, \mathrm{d}y \qquad (3)$$

where I_L is the vertically integrated value at point L; ϕ_L (x, y) is the linear basis function, assuming the value of 1 at L and 0 at the other two points.

A useful formula for the integration of linear basis functions over a triangular area is shown in equation (4) (Zienkiewicz *et al.*, 1977):

$$\iint_{\Omega_p} \phi_1^i \phi_2^j \phi_3^k \, \mathrm{d}x \, \mathrm{d}y = \frac{2 A_t \, i! \, j! \, k!}{(i + j + k + 2)!} \qquad (4)$$

where A_t is the area of the triangular subdomain.

Equation (3) was solved by expanding the polynominals and sequentially applying equation (4), then summing the N triangular subdomains until total integral is computed (Garabedian *et al.*, 1991).

For the sake of understanding the movement of the centre of mass of the Li^+ plume across the Fe^0 PRB, the area across the reactive barrier shown in Fig. 4(a) (i.e. x, 4–10.5 m; y, 4.24–7.64 m) was divided into four domains (Ω) for subsequent numerical integration. The four domains included the upgradient and downgradient locations as well as upgradient and downgradient parts of the PRB containing 10 and 16 triangular subdomains (Ω_p), respectively. Numerical integration of the Li^+ distribution in these four domains were achieved with the program (Momgen) provided by Garabedian *et al.* (1991). By finding out the time for the centre of mass of the Li^+ plume reaching the domains, the v_x across each domain can be simply computed by dividing the travelling distance with the travelling time. Table 2 shows the results of zeroth and first moments analysis in these four domains in which the maximum mass, arrival time and location of the centre of mass were included.

Table 2 Results of zeroth and first moments analysis for the Li^+ plume in the upgradient location, upgradient part, downgradient part and downgradient location of the PRB.

Domains	†Max. mass of Li^+ (g)	Time to reach max. mass of Li^+ (d)	‡$\bar{x}$ (m)	‡$\bar{y}$ (m)	‡$\bar{z}$ (masl)	§Travelling distance (m)
Upgradient location	23.74	43	6.70	7.14	17.96	2.43
Upgradient part	18.69	101	6.62	6.32	17.26	3.52
Downgradient part	23.42	105	5.92	5.51	17.03	4.61
Downgradient location	21.17	105	5.52	4.80	16.76	5.47

† zeroth moment analysis; ‡ first moment analysis–location of the centre of mass;
§ travelling distance $= \Sigma_{i=1}^{4}\sqrt{(\bar{x}_i-\bar{x}_{i-1})^2+(\bar{y}_i-\bar{y}_{i-1})^2+(\bar{z}_i-\bar{z}_{i-1})^2}$, $x_0 = 7.25$, $y_0 = 7.64$, $z_0 = 20.28$.

Regarding the results of the zeroth moment analysis, the maximum mass of Li^+ passing through the upgradient location, upgradient part, downgradient part and downgradient location of the Fe^0 PRB were similar with the percentage difference between 2.69 and 14.09%. As a result, it is highly expected that there was no loss of Li^+ across the Fe^0 PRB, thereby further substantiating the absence of bypassing flow and the nonreactivity of Li^+ within the Fe^0 packed medium. Based upon the results of the first moment analysis, the centre of mass of the Li^+ plume reaching the upgradient gravel pack on Day 43 can be examined. However, it took nearly 58 days to flow into the upgradient part of the reactive barrier corresponding to a v_x of about 6.86 m year^{-1}. Within the PRB, the Li^+ plume only took about 4 days to pass through it, which corresponded to a v_x of 99.5 m year^{-1}. However, because of low sampling frequency after Day 105, the required time for the Li^+ plume travelling from the downgradient part to downgradient location of the Fe^0 PRB could not be computed. The extreme low of v_x within the upgradient gravel pack is probably due to the effect of disturbance exerted by the clogged upgradient interface of the upper part of the Fe^0 PRB, which has been verified from the vertical distribution of the Li^+ plume (Fig. 4(b)). This disturbance effect consequently prevented the Li^+ plume from going into the preferential path readily, thereby significantly increasing its residence time within the upgradient gravel pack (Douglas *et al.*, 1995).

Consistent with the phenomena observed in Fig. 4(a) and (b), the results from the first moment analysis could also indicate the lateral displacement and significant slump of the Li^+ plume in the immediate vicinity of the front edge of the reactive barrier. From the planar view of observation, the location of the centre of mass of the Li^+ plume in the upgradient location and upgradient part of the reactive barrier were about 0.55 and 0.63 m apart from the southwest side of the barrier's centreline (x = 7.25 m), respectively. The deviation of the Li^+ flow path from the centreline was about 1.33 m in the downgradient part and 1.73 m in the downgradient location of the reactive barrier, thereby further verifying the preferential flow across the Fe^0 PRB. A significant slump of the centre of mass of the Li^+ plume could also be observed in the upgradient gravel pack in which the centre of mass dropped from a depth of 17.96 to 17.26 m a.s.l. After entering the Fe^0 PRB, the altitude of the centre of mass slightly dropped from about 17.26 m a.s.l. in the upgradient part of the reactive barrier to approx. 17.03 m a.s.l. in the downgradient part of the Fe^0 PRB.

CONCLUSIONS

Ten months monitoring of the movement of the injected Li^+ plume across the funnel-and-gate Fe^0 PRB installed at Vapokon site, Denmark, illustrated that the Li^+ plume travelled preferentially across the PRB. From the planar view of observation, the Li^+ plume was examined moving exactly along the centreline of the PRB in the location just downgradient of the injection wells. However, in the immediate vicinity of the front edge of the Fe^0 PRB, there was a pronounced lateral change of the flow path. Afterwards, the Li^+ plume travelled preferentially across the Fe^0 PRB. Mineral precipitation within the Fe^0 PRB is believed to be the main culprit causing the preferential flow since there was significant diminution in the concentrations of Ca^{2+}, Fe_T, TAL and SO_4^{2-} in the groundwater across the PRB. Comparison of the water table contour in the Vapokon site obtained in March 2000 and January 2003 manifested formation of a low permeability zone within the PRB. Based upon the results of zeroth moment analysis, the similarity of the maximum mass of Li^+ passing through the upgradient location, upgradient part, downgradient part and downgradient location of the reactive barrier was observed. Besides, first moment analysis of the Li^+ distribution indicated a v_x of 99.5 m year^{-1} within the PRB. The v_x in the immediate vicinity of the front edge of the Fe^0 PRB, however, was only about 6.86 m year^{-1} probably owing to the effect of disturbance exerted by the clogged upgradient interface of the upper part of Fe^0 PRB.

REFERENCES

Aris, R. (1956) On the dispersion of a solute in a fluid flowing through a tube. *Proc. R. Soc. London,* Ser. A, **235**, 67–78.

Broholm, M. M., Rügge, K., Tuxen, N., Højberg, A. L., Mosbæk, H. & Bjerg, P. L. (2001) Fate of herbicides in a shallow aerobic aquifer: a continuous field injection experiment (Vejen, Denmark). *Water Resour. Res.* **37**(12), 3163–3176.

Douglas, J. F., Gasiorek, J. M. & Swaffield, J. A. (1995) *Fluid Mechanics*, 3rd edn. Longman Singapore Publishers (Pte) Ltd, Singapore.

Focht, R. M., Vogan, J. L. & O'Hannesin, S. F. (1997) Hydraulic studies of in-situ permeable reactiver barriers. In: *Containment Technology Conference and Exhibition* (Proc. Int. Conf, 9–12 February, St Petersburgh, Florida, USA).

Garabedian, S. P., LeBlanc, D. R., Gelhar, L. W. & Celia, M. A. (1991) Large-scale natural gradient tracer test in sand and gravel, Cape Cod, Massachusetts 2. Analysis of spatial moments for a nonreactive tracer. *Water Resour. Res.* **27**(5), 911–924.

Gillham, R. W. & O'Hannesin, S. F. (1994) Enhanced degradation of halogenated aliphatics by zero-valent iron. *Ground Water* **32**(6), 958–967.

Lai, C. K. & Lo, I. M. C. (2002) Bench-scale study of the effects of seepage velocity on the dechlorination of TCE and PCE by zero-valent iron. In: *Environmental Geotechnology* (Proc. sixth Int. Symp., 2–5 July, Seoul, Korea).

LeBlanc, D. R., Garabedian, S. P., Hess, K. M., Gelhar, R. D., Stollenwerk, Q. K. G. & Wood, W. W. (1991) Large-scale natural gradient tracer test in sand and gravel, Cape Cod, Massachusetts, 1. Experimental design and observed tracer movement. *Water Resour. Res.* **27**(5), 895–910.

Lo, I. M. C., Lai, C. K. & Kjeldsen, P. (2003) Performance of permeable reactive barrier on remedying aliphatic chlorinated organic contaminated groundwater. In: *Contaminated Soils and Groundwater in Arid Countries* (Proc. Int. Conf., 20–23 January 20–23, Muscat, Oman).

Mackay, D. M., Freyberg, D. L., Roberts, P. V. & Cherry, J. A. (1986) A natural gradient experiment on solute transport in a sand aquifer, 1. Approach and overview of plume movement. *Water Resour. Res.* **22**(13), 2017–2029.

Puls, R. W., Blowes, D. W. & Gillham, R. W. (1999) Long-term performance monitoring for a permeable reactive barrier at the US Coast Guard Support Center, Elizabeth City, North Carolina. *J. Hazard. Mater.* **68**(1), 109–124.

Reynolds, T., Gillham, R. W. & O'Hannesin, S. F. (2002) *In Situ* monitoring of the ten-year old PRB at CFB Borden, Ontario. In: *Summary of the Remediation Technologies Development Forum Permeable Reactive Barriers Action Team Meeting* (Capital Hilton, Washington, DC, USA).

Scherer, M. M., Richter, S., Valentine, R. L. & Alvarez., P. J. J. (2000) Chemistry and microbiology of permeable reactive barriers for *in situ* groundwater clean up. *Crit. Rev. Envir. Sci. Technol.* **30**(3), 363–411.

United States Environmental Protection Agency, Office of Research and Development (1998) *Permeable Reactive Barrier Technologies for Contaminant Remediation.* EPA/600/R-98/125. US EPA, Washington, DC 20460, USA.

United States Environmental Protection Agency, Office of Solid Waste and Emergency Response Technology Innovation Office. 2002. *Field Application of In Situ Remediation Technologies: Permeable Reactive Barriers*, 68-W-00-084. US EPA, Washington, DC 20460, USA.

Zienkiewicz, O. C. (1977) *The Finite Element Method,* 3rd edn. McGraw-Hill, New York, USA.

Permeable Reactive Barriers (Proceedings of the International Symposium held at Belfast, Northern Ireland, March 2004). IAHS Publ. 298, 2005.

Quantification of the effects of pore clogging in potential PRB substrates using image analysis

JANTHAWAN WANTANAPHONG, SACHA JON MOONEY & ELIZABETH HELEN BAILEY

Environmental Science, School of Biosciences, University of Nottingham, Nottingham NG7 2RD, UK

sacha.mooney@nottingham.ac.uk

Abstract Permeable reactive barriers (PRBs) have demonstrated a promising approach for groundwater remediation. However, one concern is the deterioration of barrier material performance due to mineral precipitation and pore clogging. This study sought to quantify the effect of pore clogging on the alteration of the physical architecture of clinoptilolite and calcified seaweed using image analysis. Results after treatment with contaminated water for 1–10 months identified a decrease in porosity from approx. 22% to approx. 15% for calcified seaweed and from approx. 22% to approx. 18% for clinoptilolite. The mean pore size of both materials was also notably decreased after the treatment. An increase in irregular shaped pores was also observed in both materials after the treatment, mostly noticeably in the bottom of columns where the contaminated water first interacted with the barrier materials. The distribution of pores within an image derived from the distance transform indicated morphological differences in material between the bottom and the top of calcified seaweed columns, which is likely to impact on their performance as barrier materials.

Key words calcified seaweed; clinoptilolite; image analysis; permeable reactive barriers; pore clogging

INTRODUCTION

The longevity of barrier substrates is of crucial importance to current research efforts concerned with the development of permeable reactive barriers (PRBs). One question yet to be fully addressed is how the mechanisms of pore clogging change the architecture of a given porous media, and subsequently, how this affects the materials hydraulic and chemical performance. Recent research has shown that applying digital image processing techniques to Scanning Electron Microscopy (SEM) derived images, it is possible to examine the structure of porous materials in a quantified manner, such as surface roughness of biomaterials (Chappard *et al.*, 2003), porosity of cement paste (Yang & Buenfeld, 2001), pore size distribution of alumina membranes (Hernández *et al.*, 1997). We have applied and extended this methodology to quantify the performance of potential PRB materials. The results of this research will assist the interpretation of how precipitates alter physical properties of barrier materials in terms of their porous architecture.

MATERIALS AND METHODS

A series of column experiments were performed to test the performance of calcified seaweed and clinoptilolite at low (0.35 cm^3 min^{-1}) and high (2.08 cm^3 min^{-1}) flow rates, with different synthetic groundwater (Table 1) and with two particle size fractions (0.1–1.0 mm and 1.0–2.0 mm). After one month (high flow-rate) and 7–10 months (low flow-rate), flow was stopped, and material collected from the bottom, middle and top of the columns using a custom designed auger. This material was then examined using SEM-EDX at several magnifications to: (a) observe morphology and identify mineral precipitates (Wantanaphong *et al.*, 2005) and to (b) quantify physical properties such as total porosity, pore size, pore shape and pore distribution using image analysis. Images of representative areas (where geometric distortion was at its minima) were acquired digitally. Following a detailed pre-processing procedure to remove artefacts, including morphological erosion, each image was analysed in terms of its porous architecture using analySIS 3.0 © (Fig. 1).

Table 1 Synthetic groundwater composition used in this study. All concentration in mg l^{-1}.

Column experiments	Pb	Cu	Zn	Cd	Fe	Mn	Ca	Na	Mg	K	SO_4	Cl	HCO_3^-	CO_3^{2-}	NO_3^-
Low flow rate															
(pH 3.5)	10	10	10	1								10.84			26.59
High flow rate															
Acidic (pH 2.5)	10	10	10	1	0.5	85	50	800	550	100	3675	1237			
Neutral (pH 7)	10	10	10	1			50	40	5	3	36	17	106	75	

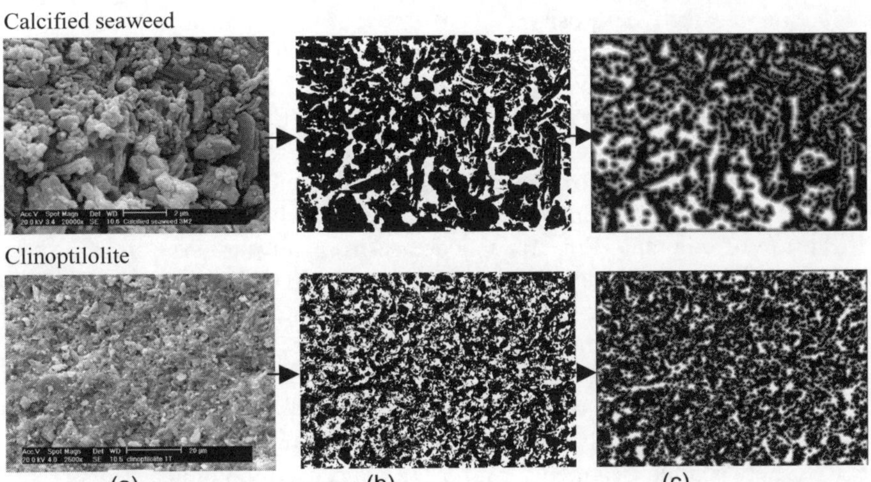

Fig 1 (a) Example of SEM images of calcified seaweed and clinoptilolite taken from columns after the experiments. (b) Binary image after the restoration, segmentation and morphological erosion of (a); the white area represents pore space and the black area represents solid space. From this, total porosity, pore size and pore shape characteristics were obtained. Images (b) were converted into distance transform images (c), which derive the distances of pore space from solid space. From this, data such as distance from pore containing half the solids (k) and pore wall length per area of image (I_A) were derived (Holden, 2001). This was used to describe the distribution of pores within an image.

Table 2 Image Analysis derived pore characteristics of calcified seaweed (CS) and clinoptilolite (ZEO).

Treatment	Sample position	Total porosity (%)		Mean pore size (μm^2)		Pore shape (% Crack & Irregular)	
		CS	ZEO	CS	ZEO	CS	ZEO
Control		21.8	22.3	0.036	0.39	88.9	88.3
Low flow rate							
Acidic groundwater (1–2 mm)	Bottom	12.2	16.2	0.018	0.43	90.5	88.9
	Top	17.2	17.4	0.028	0.47	88.3	91.9
Acidic groundwater (0.1–1 mm)	Bottom	13.7	Ongoing	0.040	Ongoing	88.6	Ongoing
(Clogged column)	Top	16.9	Ongoing	0.035	Ongoing	87.2	Ongoing
High flow rate							
Acidic groundwater (1–2 mm)	Bottom	14.1	17.0	0.026	0.29	91.4	89.9
	Top	18.0	19.8	0.036	0.38	87.1	88.6
Neutral groundwater (1–2 mm)	Bottom	14.9	15.4	0.031	0.25	89.5	89.2
	Top	16.9	18.5	0.040	0.33	87.1	88.1

RESULTS

Total porosity of both materials decreased considerably after the contaminated water treatment (Table 2). Porosity decreased from approx. 22% to approx. 15% (average values from all treatments) for calcified seaweed and from approx. 22% to approx. 18% (average values from all treatments) for clinoptilolite. It was hypothesized that total porosity at the bottom of a column, where the first treatment reaction occurred, would be less than at the column top. As expected, porosity reduction at the bottom was higher than at the top of columns in both materials. Porosity values as low as 13.7% were recorded in the bottom of a column that clogged, suggesting a significant reduction in effective porosity.

Pore size was anticipated to decrease after the water treatment in time with a reduction in total porosity. Although this was the most common occurrence, it was not always observed with the mean pore size after the treatment greater than the control in some cases. This may be explained by the comparatively small experimental sample size and heterogeneity of the materials examined. At the bottom of columns, an increased number of smaller pores was expected with a greater number of larger pores at the top of columns (Fig. 2). A decrease in mean pore size after the treatment was observed in both materials, ranging from approx. 0.029 μm^2 (at the bottom) to approx. 0.035 μm^2 (at the top) in the calcified seaweed, and approx. 0.32 μm^2 (at the bottom) to approx. 0.39 μm^2 (at the top) in the clinoptilolite.

Pore shape was postulated to be altered after the treatment with an increase in irregular shaped pores compared to the control due to the mechanisms of pore clogging, most noticeably at the bottom of a column. Results showed little difference existed between the treatments but for the majority of columns, a higher % of irregular pores was recorded at the bottom. The exception was the clinoptilolite low flow rate column which suggested the predominant removal mechanism was ion-exchange, rather than precipitation.

Distribution of pores was derived using the distance transform, which yielded data including the distance from a pore containing half the solids (K) and pore wall length per area of image (I_A). These are useful indicators of the distribution of pores within an image and gross structural architecture (Holden, 2001). A high K value indicates a low

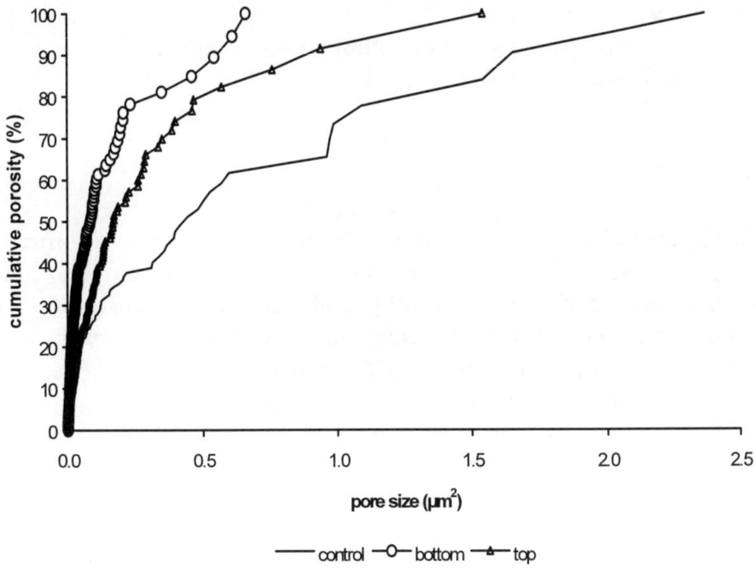

Fig. 2 Pore size distribution of calcified seaweed after reaction with acidic minewater (high flow rate), × 20 000 magnification.

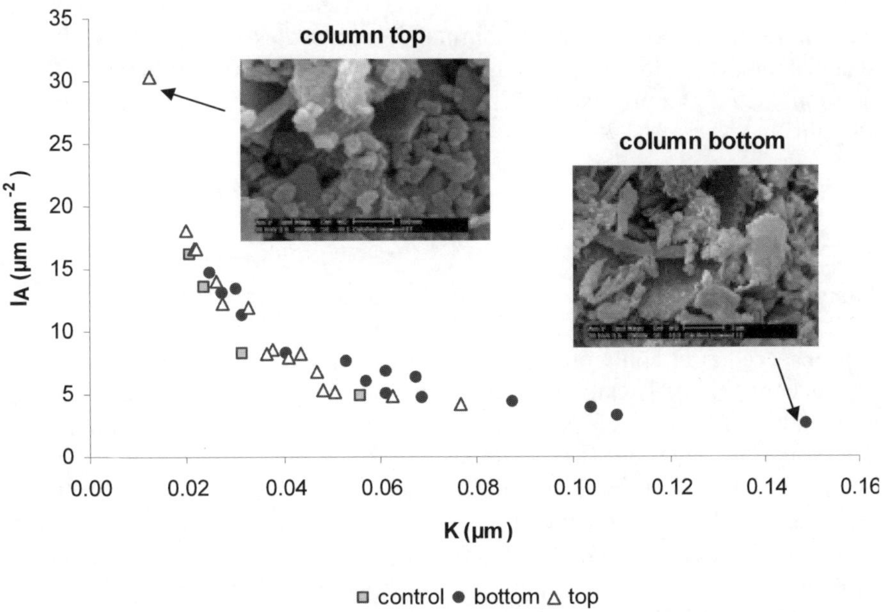

Fig. 3 Distribution of pores derived from the distance transform illustrating the difference between positions of the calcified seaweed column. A high *K* value (low porosity) was recorded in the bottom of a column.

porosity, whereas a high I_A value suggests a high porosity and surface roughness. It was expected that after treatment, especially at the bottom of a column, K would increase, whereas I_A would decrease. In general, a reduction of pore space was recorded at the bottom of a column, although this was more noticeable in the calcified seaweed (Fig. 3) than in the clinoptilolite .

DISCUSSION

Pore characteristics derived from image analysis suggested a significant change in the physical properties of calcified seaweed and clinoptilolite after the contaminated groundwater treatment. It was hypothesized that after the reaction of a material with acidic and neutral contaminated water, a decrease in total porosity and pore size and changes in pore shape, particularly at column bottom, would be expected. Results of this study revealed porosity or permeability reduction (average values after treatment compared to the control) in calcified seaweed (approx. 29%) was greater than in clinoptilolite (approx. 22%). This may due to the metal removal mechanism by calcified seaweed being precipitation, which would be likely to reduce the pore space of calcified seaweed as a consequence of the mechanism. In contrast, ion exchange is the major metal removal mechanism for clinoptilolite (Inglezakis *et al.*, 2002, 2003), which can remove metals for longer periods with no significant change in permeability or pore clogging. The morphological evidence from EDX (see Wantanaphong *et al.*, 2005) showed significant mineral precipitates (Pb, Cu and Zn carbonates) in calcified seaweed after the contaminated groundwater treatment, with no precipitates found in the clinoptilolite. It is likely that mineral precipitates resulted in a clogging of pore space. However, clogging was not recorded in a similar column with a large size fraction (1.0–2.0 mm), suggesting that particle size was a likely trigger of pore clogging.

CONCLUSIONS

Image analysis provided a means for the examination of the porous architecture of PRB materials in a novel yet quantified manner. A considerable reduction in porosity was observed after contaminated water treatment, with the greatest changes in the zone where contaminated water first interacts with the barrier material. The results were more noticeable in calcified seaweed than in clinoptilolite. The image analysis was supported by EDX results (Wantanaphong *et al.*, 2005), which showed mineral precipitates readily formed in calcified seaweed columns. These results have several important implications for the use of such materials as long term *in situ* barriers, given total effective porosity was reduced by almost a quarter within less than a year. Further research to examine the benefits of architectural quantification to predict barrier longevity is required.

REFERENCES

Chappard, D., Degasne, I., Huŕe, G., Legrand, E., Audran, M. & Basle, M. F. (2003) Image analysis measurements of roughness by texture and fractal analysis correlate with contact profilometry. *Biomater.* **24**, 1399–1407.

Hernández, A., Calvo, J. I., Prádanos, P., Palacio, L., Rodríguez, M. L. & De Saja, J. A. (1997) Surface structure of microporous membranes by computerized SEM image analysis applied to Anopore filters. *J. Membr. Sci.* **137**, 89–97.

Holden, N. M. (2001) Description and classification of soil structure using distance transform data. *Eur. J. Soil Sci.* **52**, 529–545.

Inglezakis, V. J., Loizidou, M. D. & Grigoropoulou, H. P. (2002) Equilibrium and kinetic ion exchange studies of Pb^{2+}, Cr^{3+}, Fe^{3+} and Cu^{2+} on natural clinoptilolite. *Water Res.* **36**, 2784–2792.

Inglezakis V. J., Loizidou M. D. & Grigoropoulou H. P. (2003) Ion exchange of Pb^{2+}, Cu^{2+}, Fe^{3+} and Cr^{3+} on natural clinoptilolite: selectivity determination and influence of acidity on metal uptake. *J. Colloid. Interface Sci.* **261**, 49–54.

Wantanaphong, J., Bailey, E. H. & Mooney, S. J. (2005) Suitability testing of selected wastes and natural substrates for Permeable Reactive Barriers (PRBs). In: *Permeable Reactive Barriers* (ed. by G. A. Boshoff & B. D. Bone) (Proc. Int. Symp., Belfast, Northern Ireland, March 2004), 110–116. IAHS Publ. 298. IAHS Press, Wallingford, UK (this volume).

Yang, R. & Buenfeld, N. R. (2001) Binary segmentation of aggregate in SEM image analysis of concrete. *Cem. Concr. Res.* **31,** 437–441.

Key word index

Author index

Calibration and Reliability in Groundwater Modelling: A Few Steps Closer to Reality

Edited by K. KOVAR & Z. HRKAL

IAHS Publ. 277 (*2003*) ISBN 1-901502-07-4; 526 + x pp; price £78.80

ModelCARE 2002

Financial sponsorship for this publication was provided by:

The International Hydrological Programme of UNESCO

Environmental Modeling Systems Inc., South Jordan, Utah, USA

Abstracts of the papers in this volume can be seen at:

IAHS AISH ***www.iahs.info***

with information about other IAHS publications and IAHS activities

Groundwater systems are arguably the most difficult earth systems to model because the systems are inaccessible yet, as they are a source of drinking water, very detailed knowledge is required. The usefulness of predictive simulations obtained by groundwater models is often hampered by the inability to indicate and quantify the reliability of model results. Uncertainty in predictions primarily stems from errors relating to the model formulation, such as:

- inadequate conceptualization of processes and interactions,
- inadequate description of the state of the system, and
- inadequate description of spatial and temporal variability.

Recent research has resulted in a variety of approaches that can be used to incorporate the information about such errors into the modelling process and to establish the level of uncertainty in model-based decision making. Increasingly, these new techniques are being applied to a variety of problems and sites.

The 68 papers selected from ModelCARE 2002 (the fourth of the ModelCARE conference series) and peer-reviewed, address:

Modelling Concepts
Stochastic Modelling
Parameter Estimation and Model Calibration
Conceptual Model Uncertainty and Scale
Reliability of Model Prediction
Comprehensive Use of Field Information
Methods and Tools for Assessment of Well-head Protection Zones

A major feature is the discussion of both deterministic and stochastic methods of investigation which are more often approached independently of one another.

Please send book orders and enquiries to:

Mrs Jill Gash
IAHS Press, Centre for Ecology and Hydrology
Wallingford, Oxfordshire OX10 8BB, UK

jilly@iahs.demon.co.uk
tel.: + 44 1491 692442
fax: + 44 1491 692448/692424